Mechanical engineering craft theory and associated studies
Volume 1

Consulting Editor

C. T. Butler

Head of Department of Mechanical and Production Engineering
Trent Polytechnic

Other titles in the Technical Education Series

CUNNINGTON: Calculations for Craft Students (C & G 500, 193, etc.)
DRINKALL: Basic Engineering Craft Studies (C & G 500)
FIRTH & VANDER WILLIGEN: Engineering Drawing Technology (C & G 293 & ONC)
HOUPT: Science for Mechanical Engineering Technicians (C & G 293)
KNIGHT & MELLOR: Mathematics for Mechanical Engineering Technicians. A Part 2 Course (C & G 293)
TITHERINGTON & RIMMER: Mechanical Engineering Science (ONC)
TITHERINGTON & RIMMER: Applied Mechanics (ONC)

Mechanical engineering craft theory and associated studies, volume 1

L. G. Brown
People's College of Further Education,
Nottingham

McGRAW-HILL

London · New York · Sydney · Toronto · Mexico · Johannesburg
Panama · Düsseldorf

Published by McGRAW-HILL Publishing Company Limited
MAIDENHEAD · BERKSHIRE · ENGLAND

07 094264 1

PRINTED AND BOUND IN GREAT BRITAIN

Preface

This book has been written primarily for students taking the Mechanical Engineering Craft Studies Course, Part 2 (City and Guilds No. 503). It will, however, prove useful for the Instrument Production Craft Studies Course (City and Guilds No. 504) and the older Mechanical Engineering Craft Practice Course (City and Guilds 193). The '500 series' of courses is designed to provide technical knowledge to complement the student's skill training in industry.

The contents of the book follow closely the first year Craft Theory and Associated Studies syllabuses published by the Council of Technical Examining Bodies, and a second volume is being actively prepared to cover the second year's work. It has been my aim to deal with each topic concisely, leaving detailed discussion and instruction to the individual teacher within the college workshops, classrooms, and laboratories. The sections contained within each chapter are intended to provide a framework for the teacher, and to present basic facts and principles to the student in a clear and simple manner.

The Introduction, dealing with SI Units, will prepare the student for their use throughout the book. Craft calculations are presented using worked examples; and a variety of questions, suitable for classwork or homework, is set at the end of each chapter. At the back of the book, I have included two specimen Test Papers to give students experience in handling the kind of questions which will be set in future examinations.

In conclusion, I would like to record my thanks to my wife and colleagues for their patience and valuable assistance in the preparation of this book.

L. G. BROWN

Contents

1. Introduction

To ensure that its inhabitants may enjoy a high standard of living, it is necessary for a country to secure a sound financial state. One method of helping to achieve this is to manufacture and sell products to other countries.

To improve her competitiveness in selling goods abroad, Great Britain is adopting an internationally agreed metric system of units. The units employed in this international system are called

SI UNITS (Système International d'Unités).

The craft student will become increasingly aware of these SI units during the course of his everyday work and throughout his college studies. The SI units he will meet during this Craft Studies course are those used for:

(a) Length
(b) Mass
(c) Force
(d) Temperature.

(a) LENGTH. The dimension between two positions is called a length.

The basic SI unit of length is the *metre* (m). One recommended sub-division of the metre is the millimetre (mm), this being 1/1000 m.

(b) MASS. The amount of matter in a body is called its *mass.*

The basic SI unit of mass is the kilogramme (kg). One recommended sub-division of the kilogramme is the gramme (g), this being 1/1000 kg.

(c) FORCE. Any cause which moves, or tends to move, a stationary body, is called a *force.* Similarly, any moving body may have its speed altered due to the application of a force.

The basic SI unit of force is the newton (N). A recommended multiple of the newton is the kilonewton (kN), this being 1000 N.

If a force of one newton is applied to a body whose mass is one kilogramme, then the body will move with an acceleration of one metre per second every second.

This definition is illustrated in Fig. 1.1, which shows how the speed of a body will change due to the application of a force of one newton.

(d) TEMPERATURE. The level of hotness, or coldness, of a body is called its temperature.

The derived SI unit of temperature is the degree Celsius (°C).

The freezing temperature of water is zero degrees Celsius (0°C).

The boiling temperature of water is one hundred degrees Celsius (100°C).

These SI units will become familiar to the student with their repeated use during the following chapters.

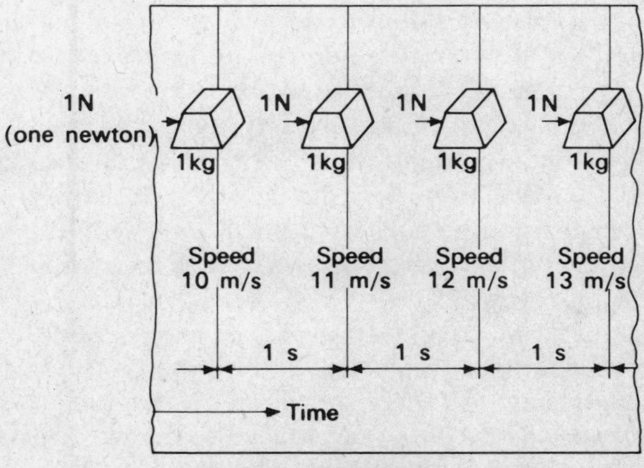

Fig. 1.1 The newton

2. Safety in the workshop

Many industrial injuries and illnesses are caused by lack of attention to safety in the workshop. The craft student should always observe the basic principles of workshop safety, both in his own interest and for the sake of those around him.

This chapter examines the following safety topics:

2.1 Protective clothing
2.2 Eye and face protection
2.3 Guarding
2.4 Workshop layout
2.5 Lifting and handling
2.6 Health hazards.

2.1 Protective clothing

Various items of protective clothing may be worn to safeguard the craftsman from injury. Protective clothing cannot guarantee absolute safety but will act as one line of defence against workshop dangers.

Overalls of some description should always be worn in the workshop, a *one-piece boiler suit* being suitable for most working conditions. A satisfactory boiler suit should be a good fit to the wearer and should not restrict natural body movements. It should fasten up high at the front by means of concealed buttons and should not have flaps on pockets or wide cuffs. A *leather apron* should be worn over the boiler suit when working with hot metals or acids, e.g., for forging, welding, or chemical cleaning processes. *Leather gauntlets* are also required during such processes, to protect the hands and forearms from burns. Suitable gloves should also be worn when handling sharp-edged components, to prevent injuries.

Protective footwear in the form of safety shoes or boots is strongly recommended for workshop wear. Such footwear, being strongly constructed and having steel toe caps, will give considerable protection from falling objects. The additional cost of such safety shoes or boots is easily justified by the length of service provided and the savings in normal shoe repairs.

Protective clothing will only be effective if worn correctly. All buttons, for example, should be kept fastened and shoe laces securely tied. In addition, the student should appreciate the danger of long hair, loose ties, torn clothing, wristlet watches and rings being caught in or between moving parts of machinery, possibly with fatal results.

2.2 Eye and face protection

The eyes and face are particularly vulnerable to injury during many workshop processes. Flying particles of metal can cause serious eye damage and therefore protection in the form of *goggles* is essential when machining metals such as brass. Similarly, goggles should always be worn during grinding processes. If, however, any irritating substance should enter the eye, this should immediately be flushed out with clean water or a recommended commercial product. The advice of an eye specialist should then be sought as soon as possible.

The bright light produced during welding processes necessitates their being viewed through darkened glass. This may be in the form of goggles, when gas welding, or a complete *faceshield*, when electric arc welding. This shield also protects the face from dangerous radiation and the splatter of molten metal.

2.3 Guarding

It is legally required that dangerous parts of machinery should be encased or *guarded*. Gear and belt drives, for example, must be securely fenced or enclosed. A wide variety of *guards* and *safety devices* is employed to protect the factory worker from dangerous parts of machinery. The management is responsible for providing such safety measures and for maintaining them in sound working order. It is, however, the responsibility of the worker to use such guards and safety devices in the correct manner. Too often, accident cases are reported where, despite the provision of safety guards, machine operators sustain serious injury due to their reluctance to use such devices.

The manufacturers of machine tools and power presses take great pains to ensure that driving and feed mechanisms are either totally enclosed or positioned beyond the reach of operators. The 'working zone' of these machines however, cannot be made safe by such measures. A lathe chuck, for example, cannot be totally enclosed or positioned remote from the operator, due to the nature of the turning process. Power presses and machine tools must therefore be fitted with special guards which provide safety whilst not hindering the working process. Typical guards and safety devices employed for such machines will therefore be considered.

(a) POWER PRESSES. Many of these machines are hand-fed, and therefore the danger of trapping the operator's hands must be totally eliminated. This is achieved by one or more of the following devices:

(i) *Interlocking guards.* These completely enclose the danger zone and are designed such that the machine will not work until the guard is in position. Similarly, interlocking guards cannot be opened while power is fed to the machine.

(ii) *Automatic guards.* These act in such a manner that when the press is set in motion, the operator's hands are removed out of danger. Such guards actually push the operator's hands to a safe position.

(iii) *Trip guards.* These automatically stop the machine if the operator's hands are endangered. A sensing device detects the presence of the operator's hand and actuates a trip mechanism which stops the machine.

(iv) *Multiple hand controls.* These depend on arranging the machine such that both of the operator's hands are required to move handles or press buttons simultaneously, before the press will operate.

(b) LATHES. The main danger of these machine tools is presented by the rotating parts and the swarf removed during cutting. Guards of the type shown in Fig. 2.1 prevent the operator from becoming caught in the revolving work and chucks or carriers, while also protecting his eyes and face from flying particles of swarf and coolant.

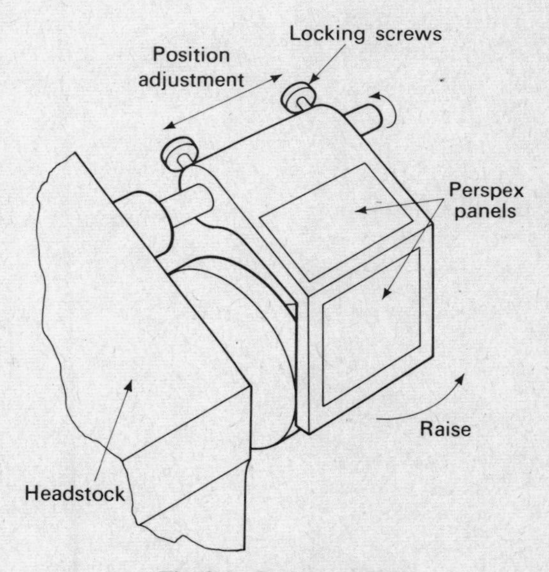

Fig. 2.1 Lathe guard

(c) MILLING MACHINES. These machines are particularly dangerous. The chief danger of milling machines lies in contact with the revolving cutter. Too often, a serious accident occurs due to an operator attempting to clear swarf while the cutter is revolving. A guard of the type shown in Fig. 2.2 will prevent such dangerous behaviour and safeguard the operator from accidental contact with the cutter.

(d) DRILLING MACHINES. These present similar dangers to those on lathes and milling machines,

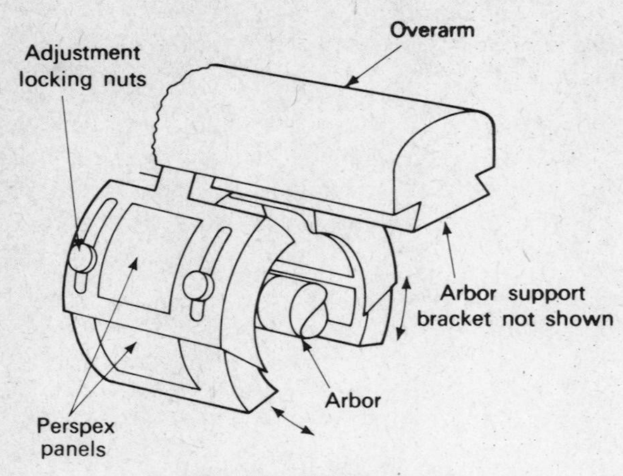

Fig. 2.2 Milling guard

and therefore drills should always be guarded. The guard employed should be telescopic or adjustable to allow for a variety of work. A typical drill guard is shown in Fig. 2.3.

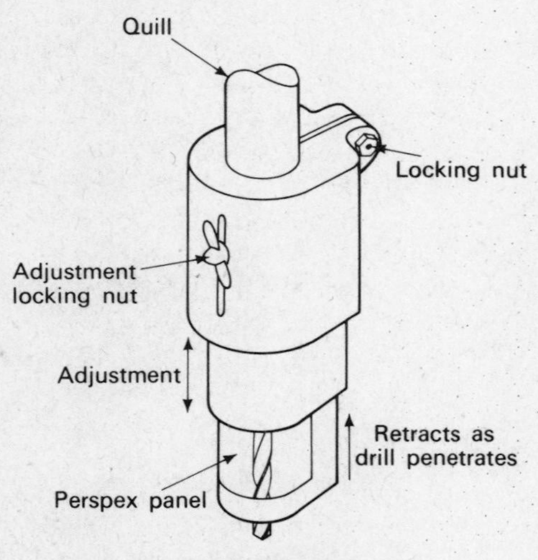

Fig. 2.3 Drill guard

2.4 Workshop layout

An engineering workshop should be conveniently laid out to suit production. For example, excessive movement of workpieces and materials between production stages is undesirable as this will increase production times and costs. The layout of a workshop must, however, provide for safety as well as convenience.

Gangways and working areas should be clearly marked out by lines painted on the workshop floor. The width of gangways should be sufficient to permit safe and easy movement of workpieces, materials and employees. Each working area should be laid out to ensure maximum safety. Machine tools must be adequately spaced apart, and arranged such that operators are positioned safely, never working back to back.

Each working area should include storage facilities for workpieces and materials. Finished work, for example, must not be allowed to clutter the workshop floor, but should be stacked safely in specially provided floorspace areas or racks. Such storage locations should be positioned to permit easy access from workshop gangways for transport equipment. The workshop floor, in addition to being kept tidy, must also be kept clean and dry. Slipping on oil patches is one obvious cause of industrial injuries which could easily be avoided.

The once common system of driving machine tools from overhead countershafts, using belts and pulleys, is rarely seen today. However where this method is still used, the parts of the driving mechanism must be guarded or positioned out of the operator's reach. Similarly, mechanisms used for feeding machine tools, e.g., bar feed mechanisms, must be safely positioned and guarded.

Every machine operator should know how to stop his machine quickly in an emergency. He may, however, be unable to do so in the event of an accident. For this reason, emergency stop buttons should be located at several convenient positions in each working area. These will enable any member of a machine shop to stop all machinery immediately in the event of an accident. It is essential that the student should observe the location and operation of such safety devices in his workshop.

2.5 Lifting and handling

Objects may be lifted manually or by mechanical means. All methods of lifting must be carried out safely to avoid accidents and injuries.

When lifting objects manually, the student should learn to take the strain with the legs by placing his feet together, bending the knees and keeping the back straight. A firm grip with the hands is essential and these should be protected by gloves when lifting sharp-edged components.

Mechanical lifting is often carried out by means of fork lift trucks, or cranes and slings. Each of these is tested periodically to establish its safe working load (S.W.L.) and this must never be exceeded. Lifting slings, in particular, must be used with caution. Two-legged slings must not be parted at an angle greater than ninety degrees, as this will reduce their safe working load. Wire-rope slings should never be sharply bent at any point or used in contact with hot metal.

Whatever means of lifting is used, vision should not be obstructed by the load and a clear route should be determined before lifting is started. Never, under any circumstances, stand beneath loads suspended by cranes.

2.6 Health hazards

Various materials and substances can harm the human body. Certain of these can cause poisoning, if they gain entry to the body. These are known as *toxic* substances. Other substances are capable of decomposing the skin tissues, thus producing a septic condition.

Many common metals, non-metallic materials, liquids and gases have toxic effects. These substances may enter the human body by various routes, including the following:

(a) THE LUNGS. Dust produced when machining alloys containing large amounts of lead, manganese, vanadium, beryllium, cadmium, and many other metal compounds can be extremely toxic. Such dusts, therefore, must not be inhaled and should be removed by suitable extraction systems, such as those fitted to surface grinders. As an added precaution, the machinist should wear a dust mask over his nose and mouth, to filter the air he breathes. Dusts produced when machining certain plastics are particularly toxic if heated. Therefore, smoking cigarettes whilst machining such materials can prove especially dangerous, as this will not only encourage inhaling toxic dust, but will also heat the dust to increase its toxic effects.

(b) THE MOUTH. Toxic substances may be taken into the body through the mouth due to smoking, eating, and drinking with dirty hands. When handling copper and aluminium materials, these actions are particularly dangerous.

(c) THE SKIN. Many toxic substances can enter the body through the skin. Lead poisoning is a particularly familiar hazard and even skin contact with lead should be avoided.

Skin disorders. Metal swarf is just one common cause of skin wounds or cuts. Germs are present everywhere in dust and dirt and these may enter skin wounds, possibly causing a septic condition. To prevent this, the wound should be cleansed immediately and a clean dressing applied. Similarly, burns should receive immediate attention. Air and germs should be excluded from burns by applying a clean dressing of gauze, lint, or fresh laundered linen. In the case of serious burns, the assistance of a doctor should be obtained immediately.

Various liquids and chemicals tend to irritate the human skin. Certain people are particularly sensitive to this irritation which may lead to severe skin inflammation called *dermatitis.* Washing hands in coolants (suds), paraffin, or petrol will remove natural protective oils from the skin and encourage irritation. Many mineral oils, used as lubricants during machining processes, contain antiseptics. Nevertheless, prolonged skin contact with such oils can cause dermatitis. Considerable protection can be obtained by applying a *barrier cream* to the hands and forearms before starting work. After finishing work, thorough washing is essential, using hot water and a special hand cleanser of the jelly type or solid soap.

Summary

The subject of workshop safety is as lengthy as it is important. Mention has been made in this chapter of only a few safety topics but these should be conscientiously remembered.

Protective clothing is a final defence against injury if other safety measures fail. Overalls, leather aprons, special footwear and gloves are typical items of protective clothing, each serving to provide protection against various dangers. Similarly, the eyes and face should be protected by goggles or shields during many workshop processes.

It is legally required that dangerous parts of machinery should be encased or *guarded.* Where complete encasing is impossible, various types of guards are used. These must protect the machine operator from injury, while not hindering the working process.

A safe workshop layout will prevent many accidents. Machines and equipment should be adequately spaced apart, and provision made for safe storage of workpieces and materials. Emergency stop buttons should be situated in convenient workshop positions, for use in the event of accidents.

Extreme care must be taken when *lifting* and *handling* components. The risk of physical strain can be reduced if a correct posture is adopted for lifting articles manually. All lifting equipment should be used correctly and within its safe working load.

Safety measures must be taken to combat the *toxic* and *septic* properties of various materials and substances. These health hazards, though perhaps less apparent, may be every bit as lethal as other workshop dangers.

Questions

1. List and describe four items of protective clothing, and explain how each of these will provide protection against injury.
2. State three workshop processes where eye and/or face protection is essential. Describe protective equipment to be worn in each case.
3. List four workshop applications of guards. Describe a suitable guard in each case.
4. Describe four possible unsafe features in workshop layouts.
5. Describe the safe method of lifting large components manually from the floor.
6. Indicate three unsafe features of the lifting arrangement shown in Fig. 2.4.

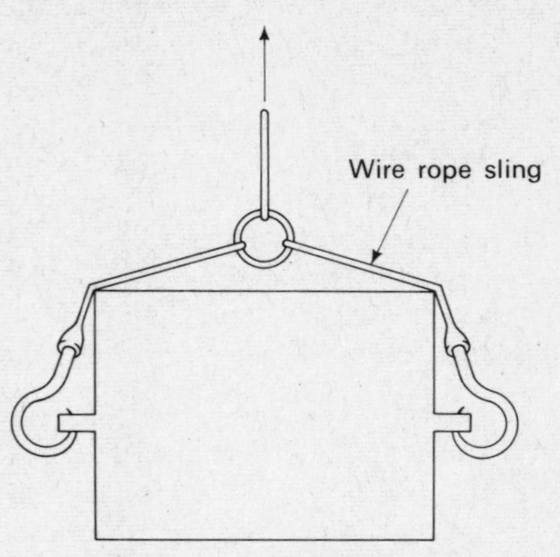

Fig. 2.4 Unsafe lifting arrangement

7. List five safety precautions to be observed when lifting by mechanical means.
8. Give three workshop examples of
   (a) toxic, and
   (b) septic hazards.

3. Dimensional control

All components are manufactured to certain dimensions. Various factors governing the choice of dimensions for components and the methods of ensuring dimensional accuracy during production, are worthy of consideration.

This chapter examines the following topics:

3.1 Standardization
3.2 Interchangeability
3.3 Gauging
3.4 The reference surface
3.5 The lead of a screw
3.6 The vernier principle.

3.1 Standardization

The reduction of several possible methods to one commonly accepted system is called *standardization*. The system resulting from standardization is called a *standard system*.

The principle of standardization has been applied widely in engineering, resulting in high-quality manufacture at reduced production costs. For example, by adopting standard forms and dimensions for features such as screw threads, it has been possible to employ mass-production methods using special tools and equipment. This has also made possible *interchangeability* (3.2), giving further economic benefits.

STANDARDS OF LENGTH. A standard is some basis upon which a standard system is founded. A standard system of linear measurement, for example, is based on a standard of length.

(a) *The Imperial Standard Yard*. This served as the Imperial standard of length from 1855 to 1960. The standard was made in the form of a bronze bar, of one inch square section, with a gold plug inserted close to each end. The distance between lines engraved on each plug was accepted as being the *Imperial Standard Yard*. This *line standard* is shown in Fig. 3.1(a).

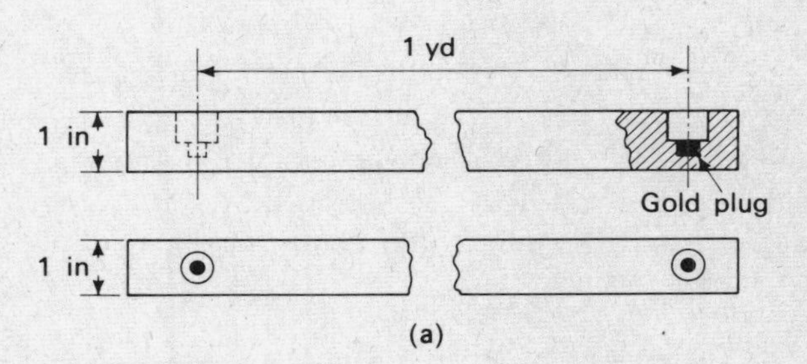

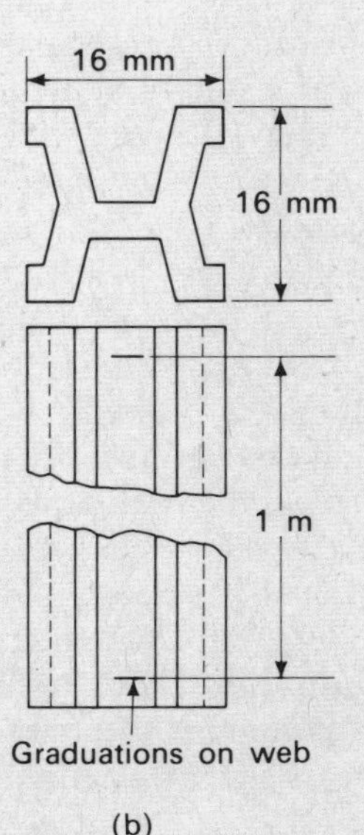

Fig. 3.1 Line standards
(a) Imperial Standard Yard
(b) International Prototype metre

(b) *The International Prototype Metre.* This served as the metric standard of length from 1875 to 1960. The standard was made from platinum–iridium alloy and was a line standard of the form shown in Fig. 3.1(b).

(c) *The Wavelength Standard.* The Imperial Standard Yard was found to be unstable, observations revealing minute shrinkage each year. Although the International Prototype Metre proved to be considerably more stable, it was internationally agreed in 1960 to define the metre in terms of a number of wavelengths of a particular radiation of light. This principle is shown simplified in Fig. 3.2.

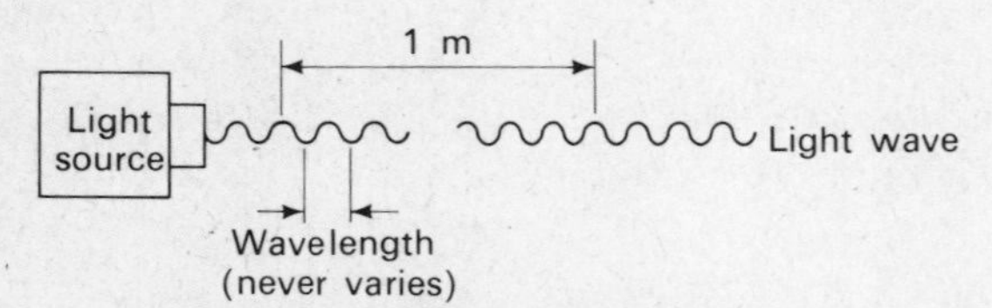

Fig. 3.2 The wavelength standard

Later the yard was defined as 0·914 4 metre and this has been the legal yard since 1964. The metric Wavelength Standard represents the SI unit of length.

STANDARD INSTITUTIONS. All major industrial countries have institutions responsible for introducing standards and standard systems.

In this country, the *British Standards Institution* (BSI) is responsible for preparing and issuing written standards. These standards apply to many branches of industry, and many domestic products manufactured to these standards may be recognized by the BSI approval symbol, the 'kite mark'.

Wherever possible, the BSI work closely with foreign standards institutions through the *International Organisation for Standardisation.* (ISO).

END STANDARDS. Although line standards have previously been employed as master standards of length, *end standards*, such as slip gauges (3.4), are generally preferred in the tool room and workshop. The main reasons for this are:

(a) It is easier to transfer a dimension from the ends of a standard, than from between two lines on its surface.

(b) Common precision measuring instruments, such as vernier calipers (3.6), depend on 'feel' for their accurate use. The accuracy of such instruments can be established by checking them against end standards.

STANDARD MEASURING TEMPERATURE. The length of any object will vary with change of temperature (4.1). For this reason, precise linear measurement should be carried out at an agreed standard temperature of 20°C.

THE STANDARDS ROOM. An instrument or gauge used in the workshop for inspecting manufactured components must have a high degree of accuracy. Such inspection devices must be checked in turn using instruments and gauges having an even higher degree of accuracy. If, for example, a diameter is turned to an accuracy of 0·05 mm, then a micrometer used for its inspection must be accurate to 0·005 mm. This micrometer must be periodically inspected using instruments or gauges measuring to an accuracy of 0·0005 mm. Such precision measurements can only be carried out under carefully controlled conditions. For this reason all primary standards of measurement used by any manufacturer, e.g. reference slip gauges (2.4), should be housed in a standards room. This will permit:

(a) Maximum care to be given to their storage and use
(b) A standard temperature of 20°C to be maintained by thermostatic control
(c) Dust-free conditions to be maintained by extreme cleanliness and filtered ventilation.

3.2 Interchangeability

Bolts are designed to fit with mating nuts and both may be manufactured by *mass-production* processes. Furthermore, any bolt chosen at random from a batch should fit with any nut similarly selected. This principle is called *interchangeability* and is applied to numerous mating components manufactured in quantity.

SELECTIVE ASSEMBLY. This is an alternative system to interchangeable assembly. When selective assembly is taken to its extreme, each component is individually measured to find ideal mating pairs for assembly.

Alternatively, components manufactured to the same given size may be separated into groups according to their degree of permitted variation from this size and suitable groups paired together for assembly.

Selective assembly is an expensive process, due mainly to the amount of measurement required.

SYSTEMS OF LIMITS AND FITS. Suitable dimensions must be chosen for all manufactured articles. Mating components, in particular, must be dimensioned to give a required type of fit. Guidance in the selection of suitable dimensions for such components is provided by *systems of limits and fits.* In order to understand the nature and use of such systems it will first be necessary to define the following basic terms:

(a) *Tolerance.* Due to the non-existence of 'perfect' machine tools and craftsmen, it is virtually impossible for any component to be produced to exact dimensions. For this reason, each dimension on a component is permitted an amount of variation. This permissible variation is called the *tolerance.*

(b) *Limits of size.* When a tolerance is applied to a dimension it has the effect of creating maximum and minimum acceptable dimensions. These are called the *maximum limit* and *minimum limit* of size, the difference between them being the tolerance.

These terms are shown applied to the width of a metal strip in Fig. 3.3(a).

(c) *Basic size.* The dimension to which a tolerance is given is called the *basic size.* The provision of a tolerance may, however, produce maximum and minimum limits remote from the basic size as shown in Fig. 3.3(b). Note that the basic size might not be an acceptable finished dimension for either of two mating parts.

(d) *Nominal size.* The size by which a dimension is commonly referred to is called the *nominal size.* The basic size and nominal size of any feature are usually identical, although this need not necessarily always be the case. A pipe thread, for example, is referred to by the diameter of the pipe's bore, this in effect becoming the nominal size of the thread.

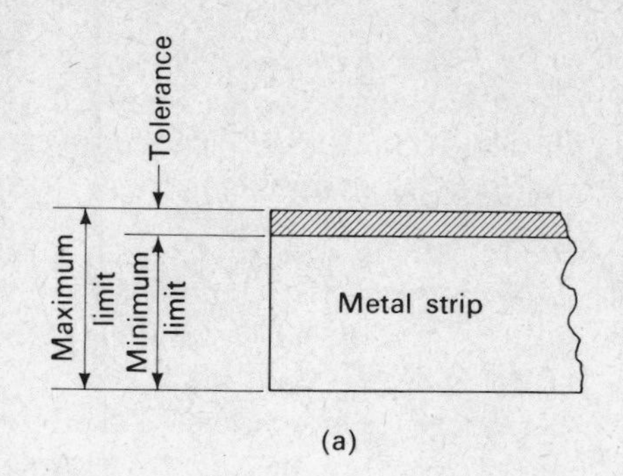

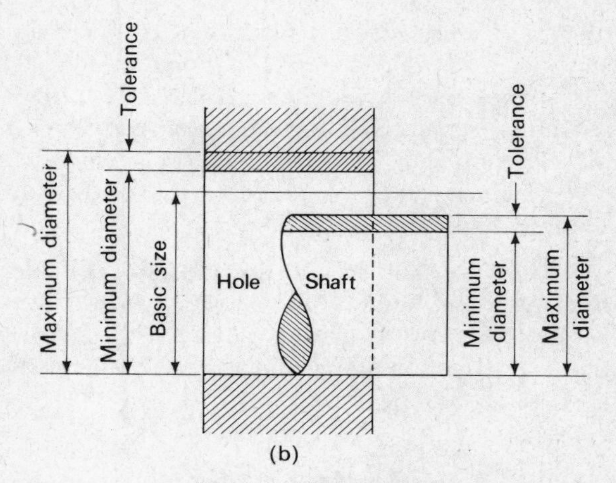

Fig. 3.3 Basic terms

(e) *Maximum metal condition.* This occurs when a shaft is manufactured to its maximum limit of size and a hole is manufactured to its minimum limit of size, i.e., the maximum amount of metal remains after machining.

(f) *Minimum metal condition.* This occurs when a shaft is manufactured to its minimum limit of size and a hole is manufactured to its maximum limit of size, i.e., the minimum amount of metal remains after machining.

Generally it is preferable for components to be manufactured to their maximum metal condition. This will ensure maximum 'life' in service and is especially important where components are subject to high rates of wear. It should be noted that if the maximum metal condition is exceeded, assembly of mating parts might well prove impossible without further machining. Machining beyond the minimum metal condition, however, will lead to the production of scrap.

(g) *Types of fit*. When parts such as shafts and holes are mated together, the fit obtained will depend on their individual sizes. Three basic types of fit are obtained between mating parts:

(i) *Clearance fit*. This is a fit which always provides clearance due to the position of the limits of size in relation to the basic size. This type of fit is shown in Fig. 3.4(a), and it should be noted that clearance is obtained (hole larger than shaft) even at the maximum metal condition.
(ii) *Interference fit*. This type of fit is shown in Fig. 3.4(b), and it should be noted that interference is always obtained (shaft larger than hole) even at the minimum metal condition.
(iii) *Transition fit*. This type of fit may provide either clearance or interference as shown in Fig. 3.4(c).

Systems of limits and fits indicate suitable tolerances and limits of size required to produce various fits between mating parts. Such information is provided in tables covering a wide range of basic sizes.

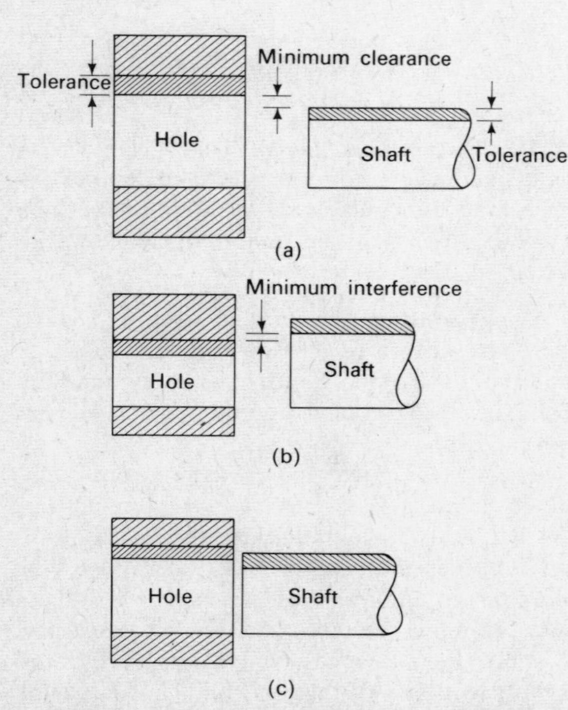

Fig. 3.4 Types of fit: (a) clearance; (b) interference; (c) transition

The British Standard System of limits and fits using metric units, is specified in BS 4500. Examples of the use of information provided in this standard are shown in chapter 4 (4.2).

3.3 Gauging

Gauges may be used to check the dimensions and form of a manufactured component. The student should already have encountered the use of templates for checking simple forms and profiles. The use of gauges for checking more complicated forms, such as those of screw threads and tapers, are discussed in Vol. 2. Therefore, at this stage, only gauges used for checking dimensions will be considered.

The dimensions of a component may be checked by two methods:

(a) *By measurement*. This involves the use of measuring instruments, such as micrometers, which will indicate the exact sizes of a component.
(b) *By gauging*. With this method *limit gauges* are used to indicate whether the dimensions of a component are within the required maximum and minimum limits of size.

Thus limit gauges do not measure a component but merely indicate whether its sizes are acceptable. The process of gauging requires little skill and can be carried out much faster than measurement. The cost of producing gauges must however be justified by sufficient quantities of components to be checked. The checking surfaces of limit gauges must be sufficiently hard to resist wear. High-carbon steel is widely used for the manufacture of gauges, these being finished by hardening and grinding. Many types of limit gauges are in use, the following being common examples:

(a) PLUG GAUGES. These are used for checking holes. Three types of plug gauge are shown in Fig. 3.5.

(i) *Double-ended plain plug gauge*. This gauge is used for checking internal dimensions. The 'GO' plug is ground to the *minimum limit* of the hole and therefore, should enter without difficulty. The 'NOT GO' plug is ground to the *maximum limit* of the hole and therefore, should not enter. Notice that the

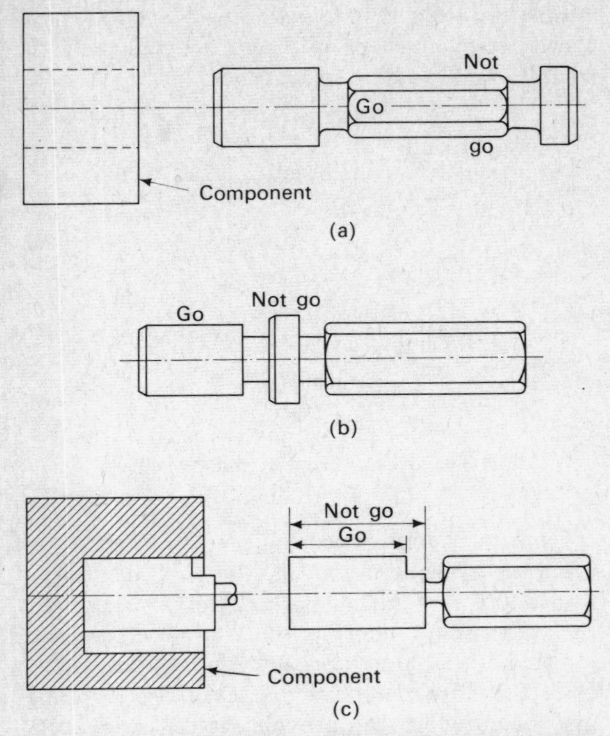

Fig. 3.5 Types of plug gauge: (a) double-ended plain plug gauge; (b) progressive plug gauge; (c) plug depth gauge

greater length of the 'GO' plug enables it to be easily distinguished from the 'NOT GO' plug.

(ii) *Progressive plug gauge*. This is used similarly to the previous type of plug gauge but does not require reversing, when checking the diameter of a hole. It cannot be used, however, for checking shallow blind holes.

(iii) *Plug depth gauge*. This gauge is used to check the depth of a hole. Its diameter is considerably smaller than that of the hole being checked, to ensure ease of entry. The depth of the hole will only be satisfactory when its surface lies between the two steps of the gauge.

(b) GAP GAUGES. These are used for checking external dimensions and diameters. Gap gauges may be either fixed or adjustable, typical examples being shown in Fig. 3.6.

(i) *Fixed-gap gauges*. These may be either double ended or progressive in form. The 'GO' gap is ground to the maximum limit of the feature being

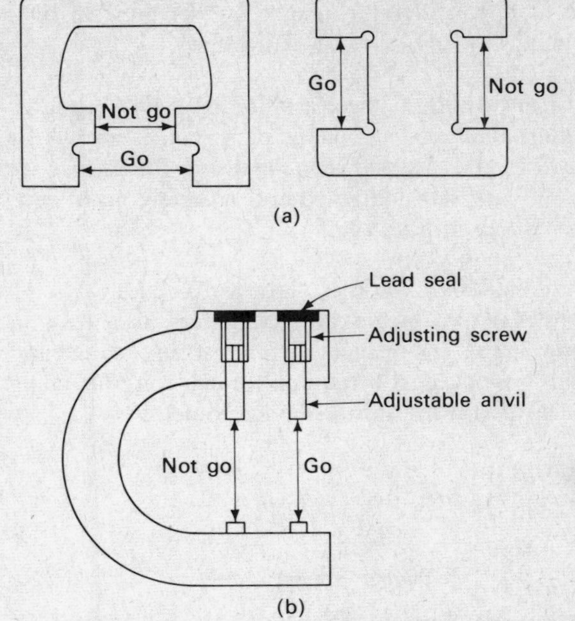

Fig. 3.6 Types of gap gauge: (a) fixed-gap gauges; (b) adjustable-gap gauge

checked, and should therefore span the feature comfortably. The 'NOT GO' gap is ground to the minimum limit of the feature, and therefore should not be able to span it.

(ii) *Adjustable-gap gauge*. This is similar to a fixed-gap gauge, but is more robust and has adjustable anvils. The anvils are set to the required gaps by means of adjusting screws which are then sealed with lead to prevent unauthorized adjustment.

(c) RING GAUGES. These are used for checking external diameters. A plain ring gauge is shown in Fig. 3.7.

This is ground to the maximum limit of an external diameter being checked and is used as a

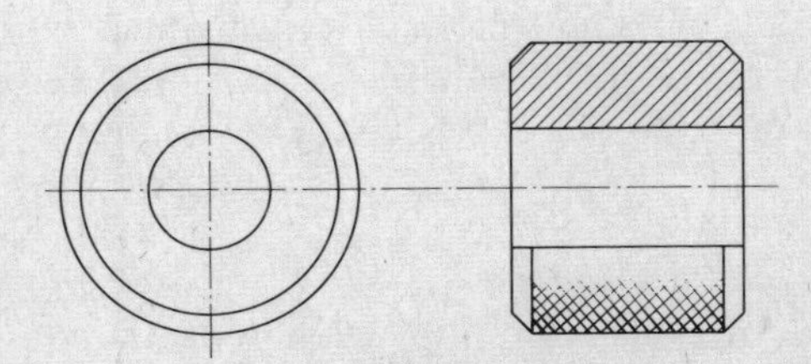

Fig. 3.7 Plain ring gauge

'GO' gauge. Such a gauge will check the diameter over a considerable length and is used in conjunction with a 'NOT GO' gap gauge.

(d) PIN GAUGES. These are used for checking large internal diameters where plug gauges would be awkward to use and very expensive. Pin gauges are steel bars with hardened and ground conical ends as shown in Fig. 3.8(a).

(e) TELESCOPIC GAUGES. These are used for checking internal dimensions and diameters as shown in Fig. 3.8(b). The retractable pin is spring-loaded and can be locked at a required dimension, this being measured using an outside micrometer.

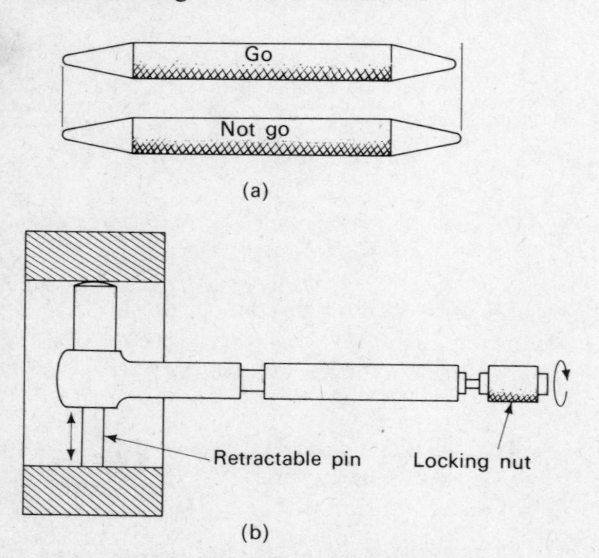

Fig. 3.8 (a) Pin gauges; (b) telescopic gauges

3.4 The reference surface

The production of accurate workpieces can be achieved only if measurements and settings are taken from reference positions or *datums*. A datum is a position from which all measurements in one direction are taken. Suitable datums must be used when marking out or setting for machining operations.

MARKING OUT FROM DATUMS. When a workpiece such as that shown in Fig. 3.9 is to be marked out for drilling, the hole centres must be established by measuring from two *datum edges* A and B. These datum edges must be filed or machined perfectly straight and square to each other.

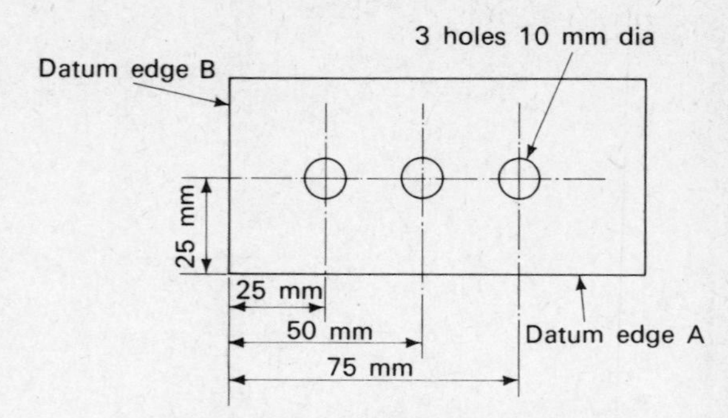

Fig. 3.9 Measurement from datum edges

Each measured dimension may be slightly inaccurate, depending on the type of measuring instrument used during marking out. For example, the error resulting from the use of a rule and scribing block might be in the order of $\frac{1}{2}$ millimetre. It is therefore necessary to select a measuring instrument which can provide the degree of accuracy required for the workpiece. It must be noted that if marking out is not undertaken from datums, errors on dimensions may accumulate as shown in Fig. 3.10.

In order to mark out the workpiece shown in Fig. 3.9 using a scribing block and rule, a surface plate or surface table must be used to provide a *reference surface* for the datum edges of the workpiece, the datum end of the rule, and the base of the scribing block as shown in Fig. 3.11.

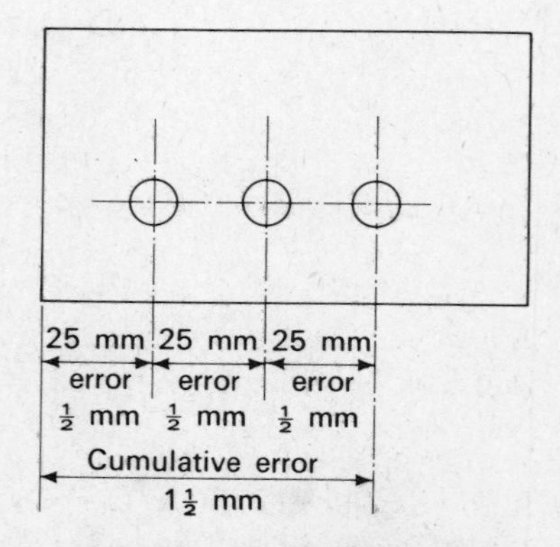

Fig. 3.10 Cumulative error

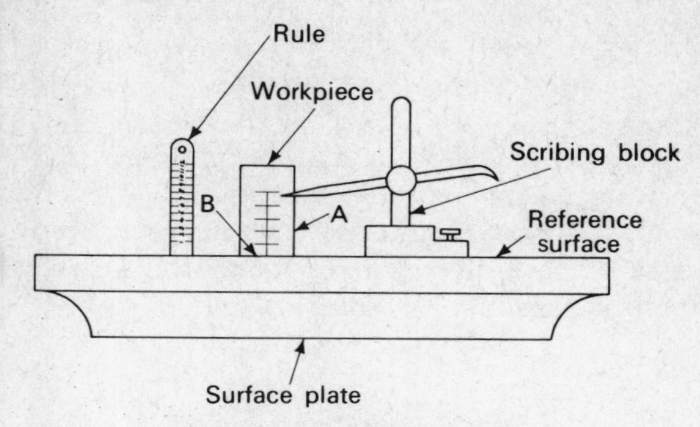

Fig. 3.11 Reference surface

When a workpiece has neither parallel nor square edges, accurate marking out can often be achieved using datums on hole centres as shown in Fig. 3.12.

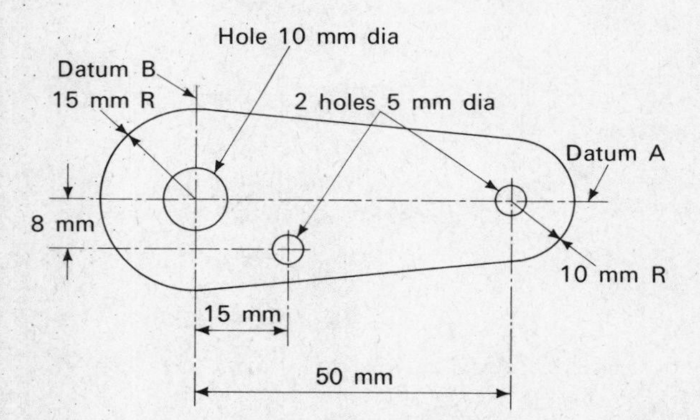

Fig. 3.12 Datums on hole centres

A scribing block, rule and surface plate could again be used for marking out, with the workpiece clamped to the vertical face of an angle plate.

SETTING FROM DATUMS. When a workpiece is set up for machining it should be located on as large a face as possible to ensure rigidity. Ideally, the locating face used should also be a datum from which dimensions are stated on an engineering drawing. Suppose, for example, the slot is to be milled in the component shown in Fig. 3.13, all other machining having been completed.

The workpiece could be clamped to the milling machine table, locating on the *datum face A*. This would ensure a rigid set-up and would enable the depth of cut to be set from the reference surface provided by the table as shown in Fig. 3.14(a). The workpiece could also be positioned correctly beneath the cutter by setting from the datum face B as shown in Fig. 3.14(b).

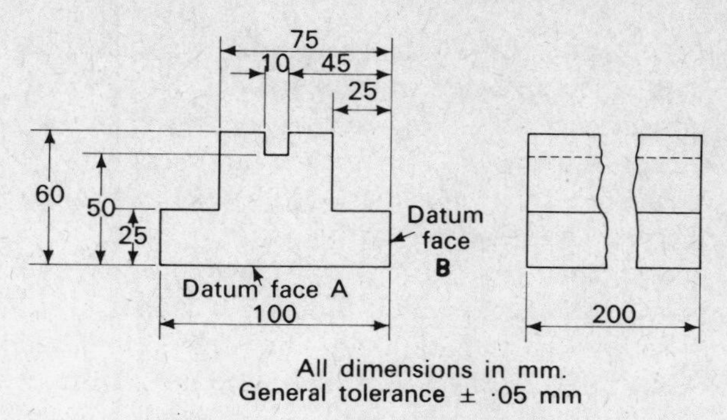

Fig. 3.13 Component to be milled

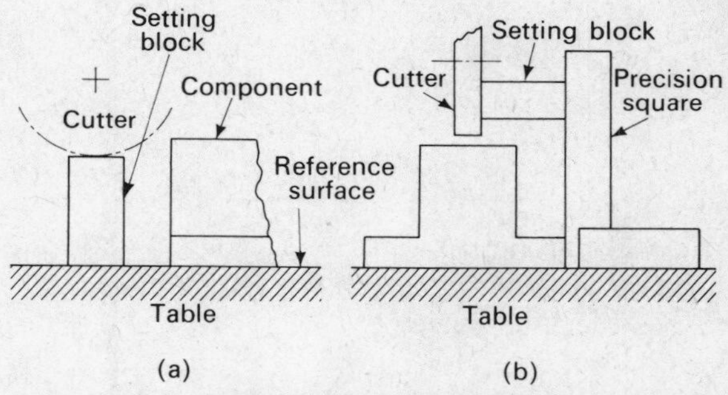

Fig. 3.14 Setting from datums

SLIP GAUGES. These are hardened and ground rectangular blocks of metal, which are lapped (11.6) to a precise dimension between two opposite faces. A *slip gauge* is, therefore, an extremely accurate *end standard*. Slip gauges of different sizes are manufactured in sets, one standard metric set containing the gauges shown in Table 3.1

Table 3.1 Set No. M78

Sizes (mm)	Number of gauges
1·01 to 1·49 by 0·01 steps	49
0·50 to 9·50 by 0·50 steps	19
10, 20, 30, 40, 50, 75, and 100	7
1·0025	1
1·005	1
1·0075	1
	78

A number of slip gauges may be combined or built up to provide an end standard for precise measurement, marking out or setting of any required dimension. Slip gauges are built up by 'wringing' them together. This involves sliding their surfaces together with a twisting action. Slip gauges correctly 'wrung' will cling together and provide their maximum degree of accuracy.

Suppose a combination of slip gauges is required to measure 27·385 mm for checking the setting of an adjustable gap gauge. The slip gauges required could be determined as in Table 3.2

Table 3.2

Gauge (mm)	Remainder
1·005	26·380
1·38	25·000
5·00	20·000
20·0	—
27·385	

Note that slip gauges are selected by starting from the right of the dimension required, i.e., from the last decimal place.

SETTING BLOCKS. These are used for setting purposes as shown in Fig. 3.14. A setting block may be manufactured from a hard-wearing material such as high-carbon steel, its length being ground to the setting dimension required.

PRECISION SQUARES AND STRAIGHT EDGES. It has previously been mentioned that two datum edges of a workpiece, used for marking out or setting purposes, must be straight and square to each other. Precision squares and straight edges may be used to check such edges when a high degree of accuracy is required. Precision straight edges may also be used to check the flatness of a reference surface such as a surface table. A typical application of a precision square for setting purposes is shown in Fig. 3.14(b).

3.5 The lead of a screw

If a screw is rotated in a fixed nut, then the screw will move axially, i.e., endwise along its axis. This principle, which is widely used for measuring instruments such as micrometers, is illustrated in Fig. 3.15(a).

Alternatively the nut may be prevented from rotating but allowed to move axially, when the screw rotates without axial movement. This principle is frequently employed for moving machine tool slides and is illustrated in Fig. 3.15(b).

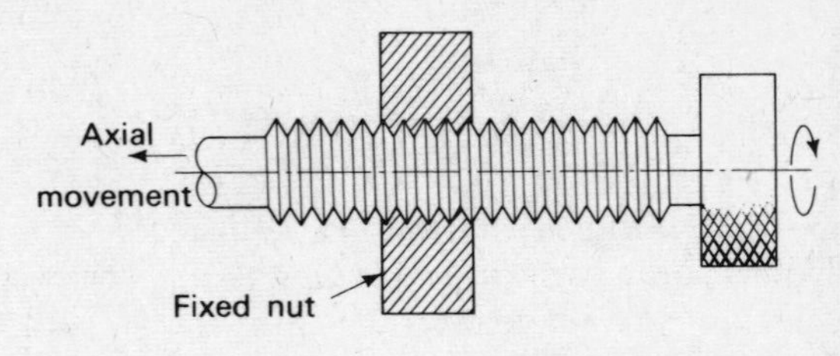

(a) Axial Screw Movement

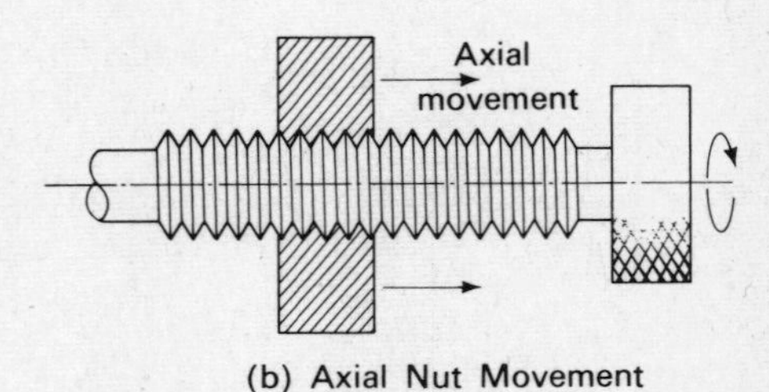

(b) Axial Nut Movement

Fig. 3.15 Axial movement: (a) axial screw movement; (b) axial nut movement

In each of the previous cases the amount of axial movement caused by one complete revolution of the screw is called the *lead of the screw*.

Both single-start and multi-start threads are used to provide axial movement in a wide variety of engineering assemblies. A single-start square thread causing axial movement of a nut is shown in Fig. 3.16(a). Note that in one revolution of the screw the nut will move axially a distance of one *pitch*, this being the spacing between threads. Therefore, for *single-start* threads,

$$\text{lead} = \text{pitch}.$$

A multi-start thread is, in effect, two or more identical threads cut side by side on one length of a screwed component. A two-start square thread causing axial movement of a nut is shown in Fig. 3.16(b). Note that, in one revolution of this screw, the nut will move axially a distance equal to twice the pitch of the thread. Therefore, for *multi-start* threads,

lead = number of starts × pitch.

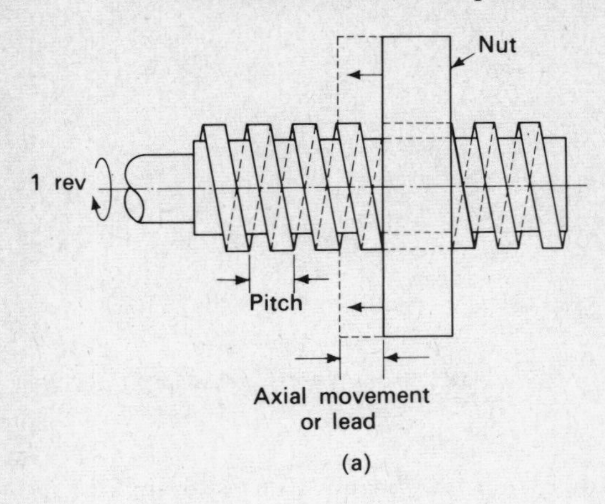

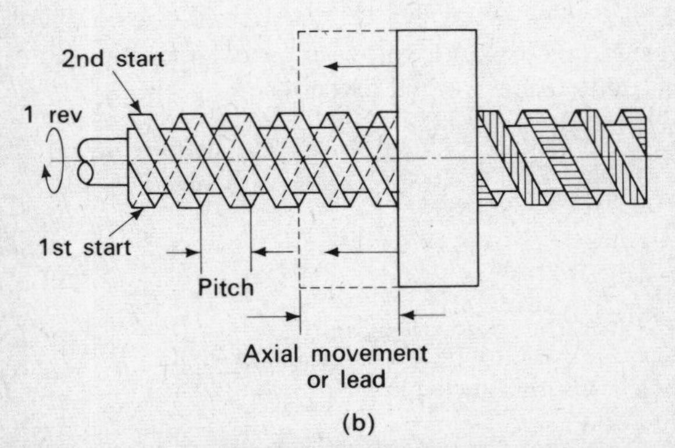

Fig. 3.16 Axial movement of nuts: (a) single-start thread; (b) multi-start thread

It should also be appreciated that the direction of axial movement will depend on two factors:

(a) The direction in which the screw is rotated.
(b) Whether a left-hand or right-hand thread is employed.

Many parts of machine tools are positioned by the axial movement caused when a screw is rotated in a nut. The headslide of a shaping machine, for example, may be moved vertically to provide a required depth of cut. This movement is obtained by rotating the operating handle controlling the headslide screw. Precise movement is made possible by the provision of an indexing dial. This indicates the vertical movement caused by rotation of the screw. The screw itself has a square thread form, and must be a good fit in the headslide nut. Clearance between the mating threads must be kept to a minimum, to avoid excessive backlash.

Calculations involving precise movements of such machine tool parts are explained in section 4.5.

3.6 The vernier principle

Measuring instruments may be used to determine the dimensions of a component. Such instruments have engraved scales graduated in convenient divisions. The *fineness* of these divisions will determine the degree of accuracy to which the scale can be read, and hence the degree of accuracy of the instrument. A metric rule, for example, is graduated in half-millimetre divisions and these represent the degree of accuracy available. To enable ease of reading and a greater degree of accuracy, many measuring instruments are provided with an additional or *vernier* scale.

The vernier principle may be explained by means of a simple example. A rule graduated in units and quarter-unit divisions is shown in Fig. 3.17(a). This rule can, therefore, be used for measuring to an accuracy of a quarter-unit.

A second scale, one unit in length and graduated in five equal divisions, may be placed against the rule as shown in Fig. 3.17(b). This second scale is called a *vernier scale*, the rule itself being the *main scale*.

The vernier scale can now be moved to the right until the graduations A and B are exactly in line as shown in Fig. 3.17(c). The amount which the vernier scale has moved will be the difference in length between the first division on the main scale and the first division on the vernier scale. This will represent the accuracy of the vernier and may be calculated as follows:

$$\begin{aligned}\text{Vernier accuracy} &= \text{1 division on main scale} - \text{1 division on vernier scale}\\ &= \tfrac{1}{4}\text{ unit} - \tfrac{1}{5}\text{ unit}\\ &= \frac{5-4}{20}\text{ unit}\end{aligned}$$

$$\therefore\ \underline{\text{vernier accuracy} = \tfrac{1}{20}\text{ unit.}}$$

This degree of accuracy is considerably better than those of the rule and vernier scales when used separately.

By constructing the scales with jaws as shown in Fig. 3.17(d) a simple vernier caliper is obtained. This is shown measuring the diameter of a bar in Fig. 3.17(e). Note that only one graduation on the vernier scale may line up with a graduation on the main scale, this being the second graduation in the setting shown. The reading is obtained as follows:

1 whole unit	= 1·0
1 quarter-unit	= 0·25
plus 2 vernier divisions = 2 × 1/20	= 0·10
Reading	= 1·35 units.

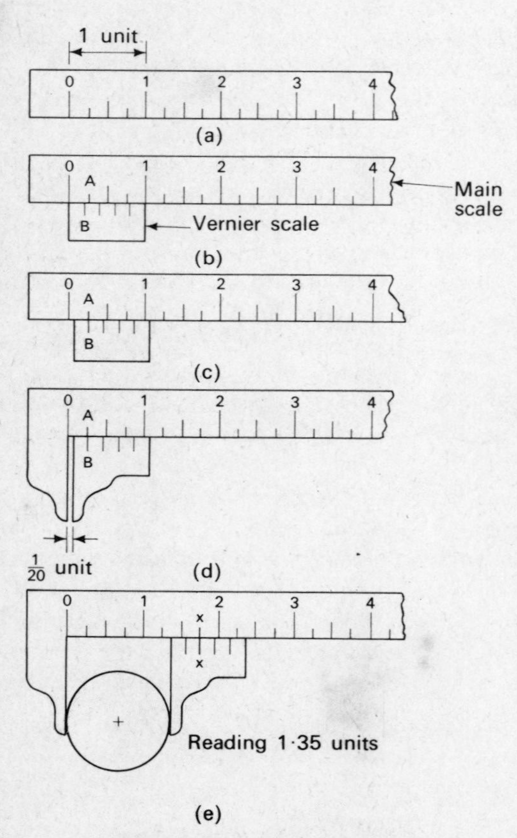

Fig. 3.17 The vernier principle

MEASUREMENT OF ANGLES. The vernier principle may also be applied to the measurement of angles; a simple vernier protractor is shown in Fig. 3.18.

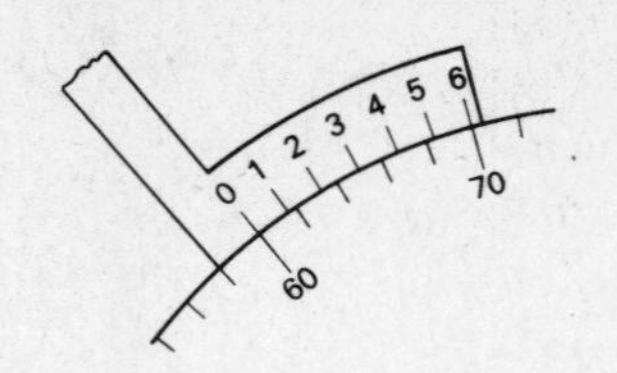

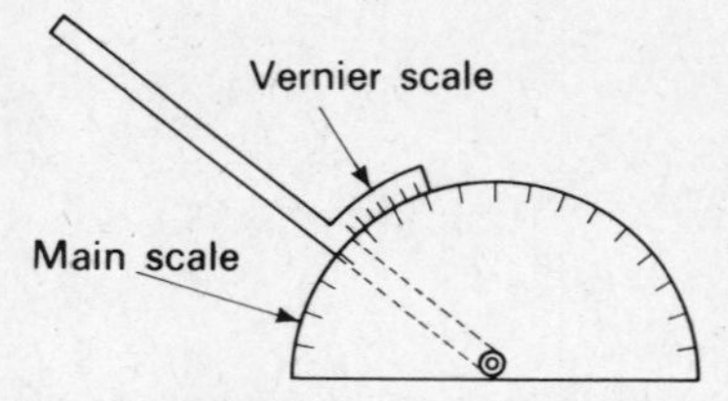

Fig. 3.18 Simple vernier protractor

Notice that the main scale is graduated in divisions of 2°, every 10° being numbered. The vernier scale has a total range of 10°, this being graduated in six equal divisions.

$$\therefore \quad \text{Vernier accuracy} = \text{1 division on main scale} - \text{1 division on vernier scale}$$

$$= 2° - 10°/6$$

$$= \frac{12 - 10°}{6} = \tfrac{2}{6}°$$

$\therefore$ vernier accuracy $= \tfrac{1}{3}°$ or $20'$.

A typical reading using this vernier protractor is shown in Fig. 3.19 and is obtained as follows:

Whole degrees	= 62°
plus 5 vernier divisions = 5 × 20′	= 1° 40′
Reading	= 63° 40′

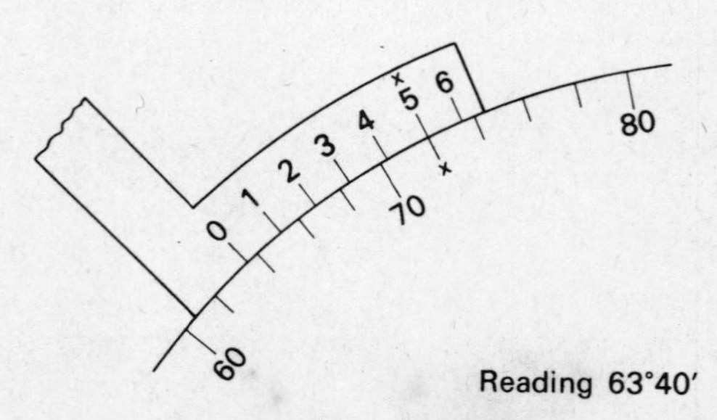

Fig. 3.19 Vernier reading

Actual vernier instruments used for engineering measurement are discussed in section 4.6.

Summary

Standardization in engineering has enabled high-quality manufacture at reduced production costs. The *British Standards Institution* (BSI) is responsible for preparing and issuing written standards in this country. This body works closely with foreign Standards Institutions through the *International Organisation for Standardisation.* (ISO).

Line standards have been used as master standards of length for many years, but *end standards* are generally preferable for workshop use. All primary standards of length within a factory should be housed in a *standards room*, where they may be used at a *standard measuring temperature* of 20°C.

Systems of *limits and fits* indicate suitable *tolerances* and *limits of size* required to produce various fits between mating parts. Such parts may be manufactured to limits which permit *interchangeability* as opposed to *selective assembly.*

The use of *gauges* for checking manufactured components eliminates the need for measuring each dimension, and considerably speeds up and simplifies the inspection process. *Plug*, *gap*, and *ring* gauges are widely used to check components for these reasons.

Measurement and marking out should always be undertaken from *datum* faces, edges or hole centres of components. The use of a *reference surface*, e.g., a surface table, will often simplify such processes. Location from datums and reference surfaces is also necessary when setting for machining operations.

Axial movement occurs when a screw is rotated in a nut. The amount of axial movement occurring during one revolution of a screw is called the *lead* of the screw.

A vernier scale used in conjunction with the main scale of a measuring instrument will greatly increase the degree of accuracy with which the instrument may be used.

Questions

1. What is *standardization* and what benefits can result from it?
2. (a) Describe the main difference between a line standard and an end standard.
   (b) Why are end standards generally preferable for workshop use?
3. Define the following terms, associated with systems of limits and fits:
   (a) Maximum limit of size
   (b) Minimum limit of size
   (c) Tolerance
   (d) Maximum metal condition
   (e) Minimum metal condition.
4. Describe, with the aid of sketches, the following types of fit:
   (a) Clearance
   (b) Interference
   (c) Transition.
5. Give one workshop example of:
   (a) An interference fit
   (b) A clearance fit.
6. What advantages does gauging have over direct measurement of finished workpieces?
7. Sketch and describe the following:
   (a) A plug gauge used for checking diameters produced to the limits 20·1/20·2 mm.
   (b) An adjustable-gap gauge.
   (c) A ring gauge used for checking shafts produced to the limits 30·1/30·2 mm dia.
8. (a) What is a datum?
   (b) What features of a component are commonly used as datums?
   (c) Explain why datums must be used when marking out, measuring or setting components for machining processes.
9. A combination of slip gauges is required to measure 45·375 mm for checking the width of a slot. Determine the slip gauges required from the 78-piece set listed in section 3.4.
10. What is meant by the *lead* of a screw.
11. (a) Calculate the accuracy of the vernier shown in Fig. 3.20.
    (b) Determine the reading shown.

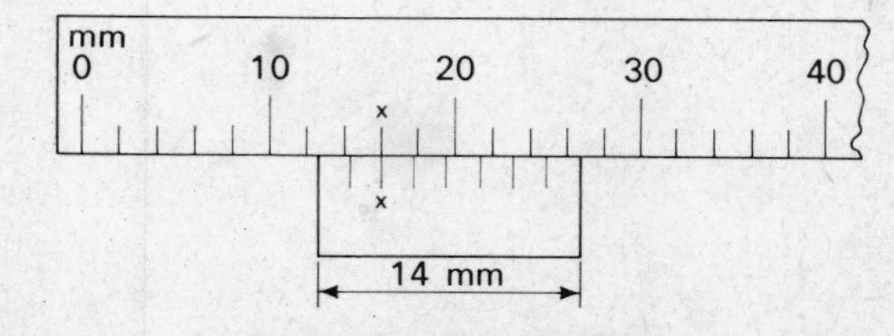

Fig. 3.20 Vernier reading

Answers

9. 1 005 mm; 1·37 mm; 3·00 mm; 40·0 mm.
11. (a) $\frac{1}{4}$ mm (b) $12\frac{1}{2}$ mm.

4. Studies associated with dimensional control

This chapter examines the following topics associated with dimensional control:

4.1 Linear expansion
4.2 Limits and fits
4.3 Geometry
4.4 Trigonometry
4.5 Pitch and lead calculations
4.6 Vernier scales.

4.1 Linear expansion

Most materials expand when heated and contract when cooled. Thus, the length of a metal object will increase due to a rise in temperature, and this is called *linear expansion.* The amount of linear expansion will depend on three factors:

(a) The original length of a metal object
(b) The amount of temperature rise
(c) The type of metal.

Linear expansion is directly proportional to the original length of a metal object. Thus, if a metal bar is cut into two pieces, one 2 m long and the other 1 m long, and these are then heated through an identical temperature rise, the linear expansion of the longer piece will be twice that of the shorter piece.

Linear expansion is also directly proportional to the amount of temperature rise. Thus, the linear expansion caused by heating a length of metal from 20°C to 60°C will be four times that caused by heating it from 20°C to 30°C.

Identical lengths of different metals expand by different amounts for the same temperature rise. Thus, each metal has its own expansion rate, and this is known as its *coefficient of linear expansion* (α). For example, the coefficient of linear expansion for cast iron is approximately 0·000 011/°C. This indicates that a length of cast iron will expand 0·000 011 m for each metre of its original length when heated from say 25°C to 26°C (one degree rise).

Table 4.1 indicates approximate coefficients of linear expansion for various engineering materials.

Table 4.1 Coefficients of linear expansion

Material	Coefficient of linear expansion/°C. (α)		
Cast Iron	0·000 011	or	$1{\cdot}1 \times 10^{-5}$
Steel	0·000 012	or	$1{\cdot}2 \times 10^{-5}$
Copper	0·000 017	or	$1{\cdot}7 \times 10^{-5}$
Brass	0·000 020	or	$2{\cdot}0 \times 10^{-5}$
Aluminium	0·000 024	or	$2{\cdot}4 \times 10^{-5}$

The three factors governing linear expansion may be combined to give the following formula:

$$\underset{(\text{m or mm})}{\text{Increase in length}} = \underset{(\text{m or mm})}{\text{original length}} \times \underset{(°\text{C})}{\text{temperature rise}} \times \alpha$$

This formula may be used to calculate linear expansion as shown in the following examples.

Example 1 A copper pipe is 3 m long at a temperature of 25°C. What increase in length will take place if the temperature of the whole pipe is raised to 225°C?

From Table 4.1, α for copper = $1{\cdot}7 \times 10^{-5}$/°C.

$$\therefore \text{Increase in length} = \text{original length} \times \text{temperature rise} \times \alpha$$

$$= 3 \times (225 - 25) \times (1{\cdot}7 \times 10^{-5})$$

$$= 3 \times 200 \times 1{\cdot}7/10^5$$

$$= 3 \times 2 \times 1{\cdot}7/10^3$$

$$= 10{\cdot}2/10^3.$$

$\therefore$ Increase in length = 0·010 2 m.

Example 2 The diameter of a steel bearing housing is to be expanded from 100 mm to 100·06 mm to simplify fitting the bearing. If the initial temperature of the housing is 5°C, calculate the temperature to which it must be raised to obtain the required expansion.

From Table 4.1, α for steel = $1{\cdot}2 \times 10^{-5}$/°C.

$$\therefore \text{Increase in length} = \text{original length} \times \text{temperature rise} \times \alpha$$

$$0{\cdot}06 = 100 \times (t - 5) \times (1{\cdot}2 \times 10^{-5}),$$

where t = temperature to which housing must be raised.

$$0{\cdot}06 = 100 \times (t - 5) \times 1{\cdot}2/10^5$$

$$\frac{0{\cdot}06 \times 10^5}{100 \times 1{\cdot}2} = t - 5$$

$$\therefore \frac{60}{1{\cdot}2} = t - 5$$

$$\therefore 50 = t - 5$$

$\therefore$ housing temperature must be raised to 55°C.

Example 3 An aluminium bar 800 mm long is mounted between centres on a lathe. During the subsequent turning operation, the temperature of the bar increases from 15°C to 25°C. Calculate the increase in length of the bar during machining.

From Table 4.1, α for aluminium = $2{\cdot}4 \times 10^{-5}$/°C.

$$\therefore \text{Increase in length} = \text{original length} \times \text{temperature rise} \times \alpha$$

$$= 800 \times (25 - 15) \times (2{\cdot}4 \times 10^{-5})$$

$$= 800 \times 10 \times 2{\cdot}4/10^5$$

$$= 8 \times 1 \times 2{\cdot}4/10^2$$

$\therefore$ Increase in length = 0·192 mm.

Note the necessity for using coolant and adjusting the tailstock during such processes to prevent the centre from being 'burnt out'.

4.2 Limits and fits

Systems of limits and fits indicate suitable limits of size for mating components to provide a required *type of fit* (3.2). Two types of systems are used for indicating such limits, these being:

(a) A hole-basis system
(b) A shaft-basis system.

(a) HOLE-BASIS SYSTEM. This is a system of limits and fits in which different clearances and interferences are obtained by mating various shafts with a single hole. This system is recommended for most engineering purposes, because a shaft may be finish machined to any desired size without difficulty. A hole, however, can be more conveniently finished using 'standard size' cutting tools such as drills and reamers. Also, plug gauges (3.3) used for checking holes, are manufactured in standard sizes.

(b) SHAFT-BASIS SYSTEM. This is a system of limits and fits in which different clearances and interferences are obtained by mating various holes with a single shaft. This system may be more convenient for selecting limits of size when a long shaft has to carry several mating parts such as bearings, pulleys, gears and couplings.

BS 4500: SPECIFICATION FOR ISO LIMITS AND FITS. This British Standard System of Limits and Fits was published in 1969. It contains charts and tables of dimensions from which may be determined suitable limits of size to provide any required fit between mating parts.

To simplify the use of this standard, various selected fits are recommended, these being suitable for most engineering purposes.

SELECTED HOLE-BASIS FITS. Various selected hole-basis fits for a 25 mm diameter shaft and hole are shown in Fig. 4.1.

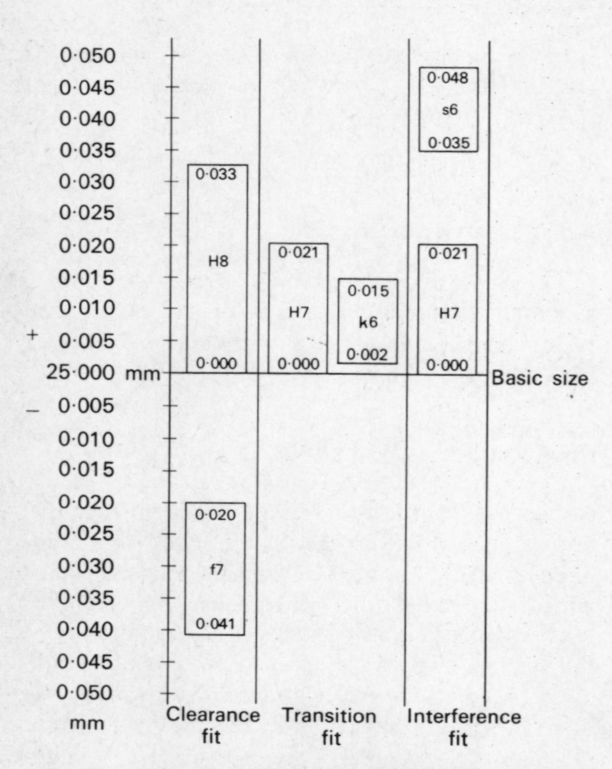

Fig. 4.1 Selected fits (hole-basis) for 25 mm diameter

The chart shown in Fig. 4.1 will appear complicated at first sight. However, it is simple to use if the following points are noted and remembered:

(a) All holes are given a code letter and number. The letter used for a hole will always be a *capital letter*, e.g., a typical hole is referred to as H8.
(b) All shafts are given a code letter and number. The letter used for a shaft will always be a *small letter*, e.g., a typical shaft is referred to as f7.
(c) The tolerance given to any hole or shaft is shown as a block. For example, a 25 mm diameter H8 hole has a tolerance of 0·033 mm and this produces the following limits of size:

$$\text{Maximum limit} = 25{\cdot}000 + 0{\cdot}033$$
$$= 25{\cdot}033 \text{ mm dia.}$$

$$\text{Minimum limit} = 25{\cdot}000 + 0{\cdot}000$$
$$= 25{\cdot}000 \text{ mm dia.}$$

Similarly a 25 mm diameter f7 shaft has a tolerance of 0·021 mm and this produces the following limits of size:

$$\text{Maximum limit} = 25{\cdot}000 - 0{\cdot}020$$
$$= 24{\cdot}980 \text{ mm dia.}$$

$$\text{Minimum limit} = 25{\cdot}000 - 0{\cdot}041$$
$$= 24{\cdot}959 \text{ mm dia.}$$

Notice that for this shaft the *basic size* of 25·000 mm diameter is not an acceptable finished dimension. For each of the H holes shown on the chart, however, the minimum limit of size is in fact the basic size, i.e., 25·000 mm diameter. This enables the use of standard cutting tools, such as 25 mm diameter drills and reamers. The tolerances provided on each of these holes allows for cutting tools producing oversize holes.

Example 1 A shaft and mating hole are to be manufactured to a nominal size of 25 mm diameter and a loose clearance fit is required.

(a) What is meant by the term *nominal size*?
(b) Calculate the minimum clearance and maximum clearance obtained using a fit of H8 f7.

(a) The *nominal size* is a stated value used to give some indication of the size of a feature. For example, a shaft manufactured to 20·00/20·05 mm diameter might be referred to as a 20 mm diameter shaft. Thus 20 mm diameter is the nominal size of this shaft. Usually (but not necessarily) the nominal size and basic size of a feature are identical.
(b) Minimum clearance = Smallest hole − Largest shaft.
Referring to Fig. 4.1:

$$\text{Smallest H8 hole} = 25{\cdot}000 \text{ mm dia.}$$

$$\text{Largest f7 shaft} = 25{\cdot}000 - 0{\cdot}020$$
$$= 24{\cdot}980 \text{ mm dia}$$

$$\therefore \quad \text{Minimum clearance} = 25{\cdot}000 - 24{\cdot}980$$
$$= \underline{0{\cdot}020 \text{ mm.}}$$

Maximum Clearance = Largest hole − Smallest shaft.

Referring to Fig. 4.1:

Largest H8 hole = 25·000 + 0·033
= 25·033 mm dia

Smallest f7 shaft = 25·000 − 0·041
= 24·959 mm dia

∴ Maximum clearance = 25·033 − 24·959
= <u>0·074 mm.</u>

Example 2 A bearing is to be a press fit in its housing. If the nominal size of the bearing is 25 mm outside diameter and an interference fit of H7 s6 is selected for the manufacture of the bearing and its housing, calculate:

(a) The minimum interference obtained
(b) The maximum interference obtained.

(a) Minimum interference = Smallest shaft − Largest hole.
Referring to Fig. 4.1:

Smallest s6 shaft = 25·000 + 0·035
= 25·035 mm dia

Largest H7 hole = 25·000 + 0·021
= 25·021 mm dia

∴ Minimum interference = 25·035 − 25·021
= <u>0·014 mm.</u>

(b) Maximum interference = Largest shaft − Smallest hole.
Referring to Fig. 4.1:

Largest s6 shaft = 25·000 + 0·048
= 25·048 mm dia

Smallest H7 hole = 25·000 mm dia

∴ Maximum interference = 25·048 − 25·000
= <u>0·048 mm.</u>

Notice that the male feature is always referred to as a shaft and the female feature as a hole for the purpose of such calculations.

SELECTED SHAFT-BASIS FITS. Various shaft-basis fits for a 60 mm diameter shaft and hole are shown in Fig. 4.2.

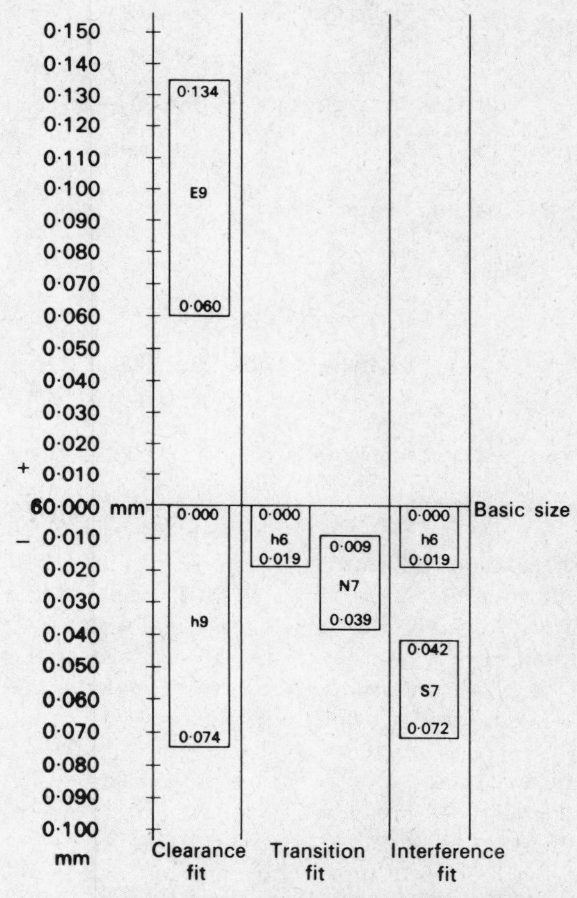

Fig. 4.2 Selected fits (shaft-basis) for 60 mm diameter

Example A gearwheel is to be a push fit on a shaft whose nominal size is 60 mm diameter. It is required that the shaft and bore of the gearwheel should be machined to a shaft-basis fit of N7 h6.

(a) What type of fit is produced?
(b) Calculate the maximum clearance obtained.
(c) Calculate the maximum interference obtained.

(a) Referring to Fig. 4.2:
N7 h6 is a *Transition fit.*
(b) Maximum clearance = Largest hole − Smallest shaft.

Referring to Fig. 4.2:

$$\text{Largest N7 hole} = 60{\cdot}000 - 0{\cdot}009$$
$$= 59{\cdot}991 \text{ mm dia}$$
$$\text{Smallest h6 shaft} = 60{\cdot}000 - 0{\cdot}019$$
$$= 59{\cdot}981 \text{ mm dia}$$
$$\therefore \quad \text{Maximum clearance} = 59{\cdot}991 - 59{\cdot}981$$
$$= \underline{0{\cdot}010 \text{ mm.}}$$

(c) Maximum interference = Largest shaft – Smallest hole.
Referring to Fig. 4.2:

$$\text{Largest h6 shaft} = 60{\cdot}000 \text{ mm dia}$$
$$\text{Smallest N7 hole} = 60{\cdot}000 - 0{\cdot}039$$
$$= 59{\cdot}961 \text{ mm dia}$$
$$\therefore \quad \text{Maximum interference} = 60{\cdot}000 - 59{\cdot}961$$
$$= \underline{0{\cdot}039 \text{ mm}}$$

UNILATERAL TOLERANCE. By referring to the charts shown in Fig. 4.1 and Fig. 4.2 it will be noted that every tolerance shown is either totally above or totally below the basic size. A tolerance of this type is called a *unilateral tolerance* (one way) and is said to produce *unilateral limits of size*.

BILATERAL TOLERANCE. This is a tolerance which extends both above and below the basic size as shown in Fig. 4.3.

The complete dimension indicated in Fig. 4.3 could be written:

$$80{\cdot}000 \begin{matrix} +\,0{\cdot}020 \\ -\,0{\cdot}015 \end{matrix} \text{ mm} \qquad \text{or} \qquad 80{\cdot}020/79{\cdot}985 \text{ mm.}$$

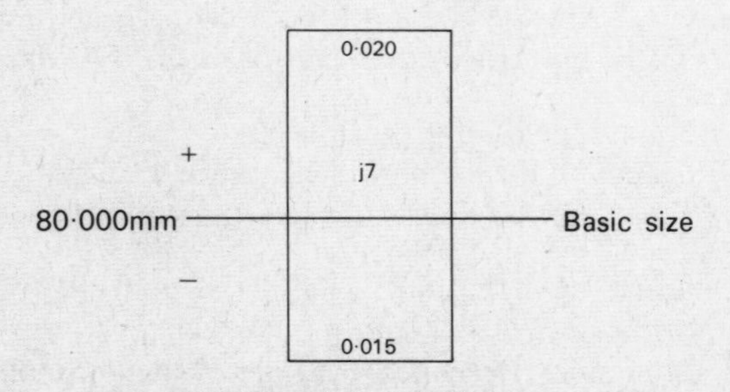

Fig. 4.3 Bilateral tolerance

Thus bilateral limits of size are obtained, the maximum limit (80·020 mm) being above the basic size and the minimum limit (79·985 mm) being below the basic size.

4.3 Geometry

A basic knowledge of geometry is required to enable many workshop calculations to be carried out. Therefore the craft student must learn the basic geometry of angles and circles.

GEOMETRY OF ANGLES. Angles may be stated either in degrees and decimal fractions of a degree, e.g., 30·24°, or in degrees, minutes and seconds, e.g., 30° 14′ 24″. It must therefore be remembered that

$$1 \text{ degree} = 60 \text{ minutes}$$

and

$$1 \text{ minute} = 60 \text{ seconds.}$$

Various common types of angle are shown in Fig. 4.4.

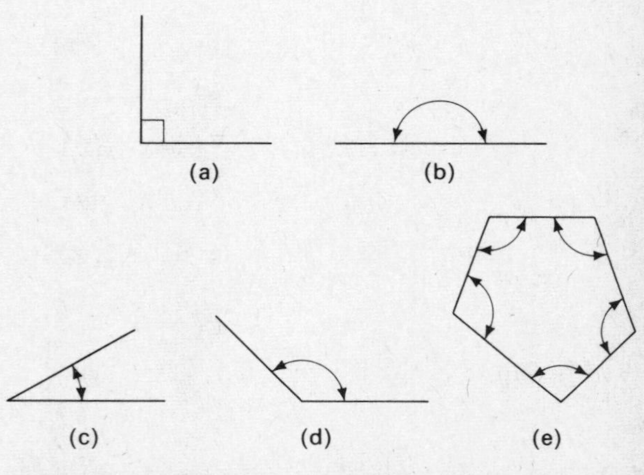

Fig. 4.4 Common types of angle: (a) right angle; (b) two right angles; (c) acute angle; (d) obtuse angle; (e) interior angles

The following points should be noted:

(a) A *right* angle = 90°
(b) Two right angles = 180° (a straight line)
(c) An *acute* angle = less than 90°
(d) An *obtuse* angle = greater than 90° but less than 180°

(e) The sum of the interior angles of any plane figure may be calculated using the formula: Sum of interior angles = $2N - 4$ right angles, where N = the number of sides of the figure.

Example Calculate the sum of the interior angles of any triangle.

$$\begin{aligned}\text{Sum of interior angles} &= 2N - 4\\ &= (2 \times 3) - 4\\ &= 6 - 4 = 2 \text{ right angles}\\ &= \underline{180^\circ}.\end{aligned}$$

When two or more straight lines intersect (cross), various related angles are produced. Typical related angles are shown in Fig. 4.5.

The following points should be noted:

(a) *Corresponding* angles are equal
(b) *Alternate* angles are equal
(c) *Opposite* angles are equal
(d) *Complementary* angles total 90°
(e) *Supplementary* angles total 180°.

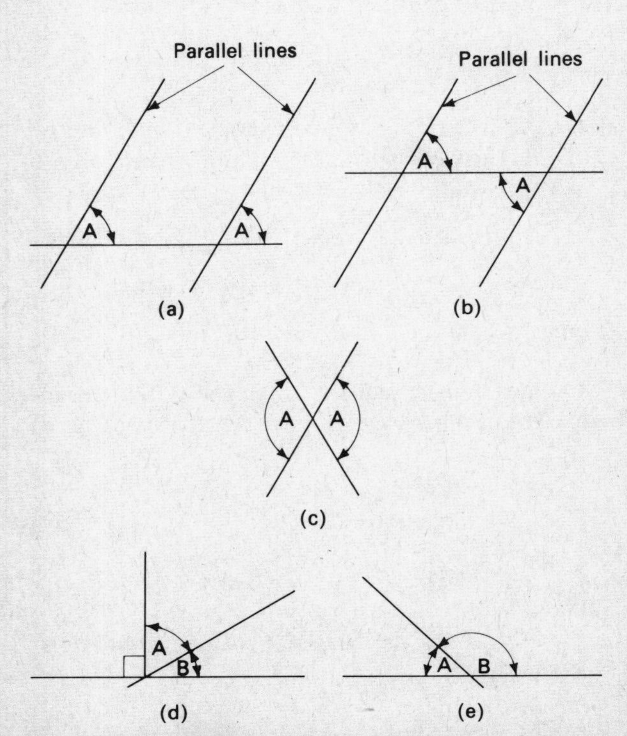

Fig. 4.5 Related angles: (a) corresponding; (b) alternate; (c) opposite; (d) complementary; (e) supplementary

The use of such relationships will be appreciated by working through the questions at the end of this chapter.

GEOMETRY OF CIRCLES. Various features of a circle, with which the student should already be familiar, are shown in Fig. 4.6.

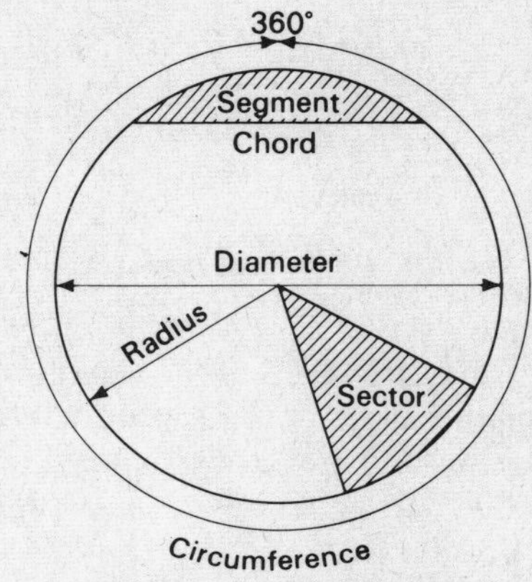

Fig. 4.6 Features of a circle

To enable certain calculations to be made, two further points must be noted:

(a) A straight line which touches the circumference of a circle at only one point is called a *tangent*, and this will be at 90° to a radius as shown in Fig. 4.7(a).
(b) When any triangle is inscribed in a semi-circle, as shown in Fig. 4.7(b), the angle opposite the diameter will be a right angle.

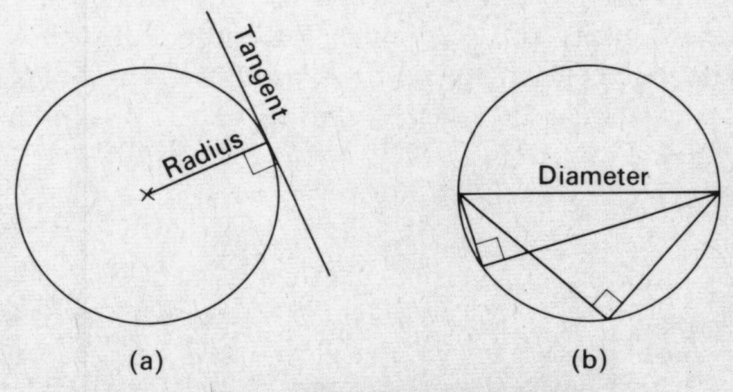

Fig. 4.7 Tangents and right angles

4.4 Trigonometry

A right-angled triangle has three interior angles, one of which equals 90°. The values of the two remaining angles will depend on the lengths of the triangle's three sides.

A right-angled triangle is shown in Fig. 4.8, each side being named according to its position relative to the angle A.

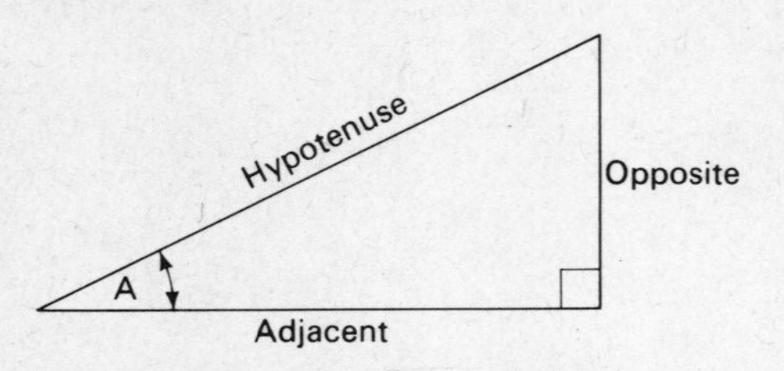

Fig. 4.8 Right-angled triangle

The sides may be taken in pairs and expressed as ratios of the angle A. These are called *trigonometrical ratios* and are known as *sine*, *cosine* and *tangent*. These three ratios are expressed as follows:

Sine ratio: $\sin A = \dfrac{\text{OPPOSITE}}{\text{HYPOTENUSE}}$

Cosine ratio: $\cos A = \dfrac{\text{ADJACENT}}{\text{HYPOTENUSE}}$

Tangent ratio: $\tan A = \dfrac{\text{OPPOSITE}}{\text{ADJACENT}}.$

The use of these ratios for workshop calculations involving right angled triangles will be demonstrated later.

USE OF TABLES. All angles have natural sine, cosine, and tangent ratios. Values for these have been calculated for angles up to 90° and listed in tables. An extract from *natural sine* tables is shown in Fig. 4.9.

Note that degrees are listed in the extreme left-hand vertical column. Each other vertical column is headed by a certain value in minutes.

Example 1 Obtain sin 2° 0′ from Fig. 4.9.

(a) Read down the extreme left-hand column to find 2°

(b) Next read across to the vertical column headed 0′ and note:

$$\sin 2^\circ\, 0' = \underline{0{\cdot}0349.}$$

Example 2 Obtain sin 3° 30′ from Fig. 4.9.

(a) Read down the extreme left-hand column to find 3°

(b) Next read across to the vertical column headed 30′ and note:

$$\sin 3^\circ\, 30' = \underline{0{\cdot}0610.}$$

Example 3 Obtain sin 4° 44′ from Fig. 4.9.

(a) Read down the extreme left-hand column to find 4°

(b) Next read across to the vertical column headed 42′ and note:

$$\sin 4^\circ\, 42' = 0{\cdot}0819$$

(c) 4° 42′ is 2′ smaller than the angle stated. Therefore read across to the *mean difference* column (ADD column) headed 2′ and note:

$$\text{Mean difference} = 6$$

$$\therefore \quad \sin 4^\circ\, 44' = 0{\cdot}0819 + 6$$

$$= \underline{0{\cdot}0825.}$$

Cosines and tangents of angles are obtained in an identical manner from natural cosine and

NATURAL SINES

Degrees	0′	6′	12′	18′	24′	30′	36′	42′	48′	54′	Mean Differences				
	0·0°	0·1°	0·2°	0·3°	0·4°	0·5°	0·6°	0·7°	0·8°	0·9°	1	2	3	4	5
0	·0000	0017	0035	0052	0070	0087	0105	0122	0140	0157	3	6	9	12	15
1	·0175	0192	0209	0227	0244	0262	0279	0297	0314	0332	3	6	9	12	15
2	·0349	0366	0384	0401	0419	0436	0454	0471	0488	0506	3	6	9	12	15
3	·0523	0541	0558	0576	0593	0610	0628	0645	0663	0680	3	6	9	12	15
4	·0698	0715	0732	0750	0767	0785	0802	0819	0837	0854	3	6	9	12	15
5	·0872	0889	0906	0924	0941	0958	0976	0993	1011	1028	3	6	9	12	14

Fig. 4.9 Extract from natural sine tables

tangent tables. However, when using tables of natural cosines, any mean difference must be subtracted to obtain the final value.

Example 4 Using tables of natural sines, cosines and tangents, check the following:

(a) Sin 68° 21′ = 0·9294
(b) Sin 17° 58′ = 0·3085
(c) Cos 39° 25′ = 0·7725
(d) Cos 83° 46′ = 0·1085
(e) Tan 9° 14′ = 0·1626
(f) Tan 49° 5′ = 1·1538
(g) The angle whose *sin* is 0·3461 = 20° 15′
(h) The angle whose *cos* is 0·6932 = 46° 7′
(j) The angle whose *tan* is 1·0643 = 46° 47′.

TRIGONOMETRY CALCULATIONS. Various calculations involve the use of trigonometry; the following examples will indicate the methods to be used.

Example 1 Calculate the length of the side marked *x* in each of the triangles shown in Fig. 4.10.

(a) The *given side* is the HYPOTENUSE.
The *required* side is OPPOSITE to the *given* angle.

$$\therefore \quad \sin 32° 31' = \frac{\text{OPPOSITE}}{\text{HYPOTENUSE}}$$

$$\therefore \quad 0{\cdot}5375 = x/5$$

$$\therefore \quad 5 \times 0{\cdot}5375 = x$$

$$\therefore \quad \underline{x = 2{\cdot}6875 \text{ m.}}$$

(b) The *given* side is the HYPOTENUSE.
The required side is ADJACENT to the *given* angle.

$$\therefore \quad \cos 22° 15' = \frac{\text{ADJACENT}}{\text{HYPOTENUSE}}$$

$$\therefore \quad 0{\cdot}9256 = x/8$$

$$\therefore \quad 8 \times 0{\cdot}9256 = x$$

$$\therefore \quad \underline{x = 7{\cdot}4048 \text{ m.}}$$

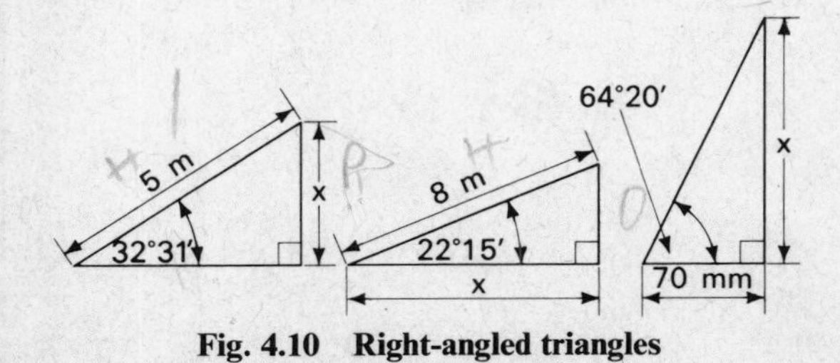

Fig. 4.10 Right-angled triangles

(c) The *given* side is ADJACENT to the *given* angle.
The *required* side is OPPOSITE to the given side.

$$\therefore \quad \tan 64° 20' = \frac{\text{OPPOSITE}}{\text{ADJACENT}}$$

$$\therefore \quad 2{\cdot}0809 = x/70$$

$$\therefore \quad 70 \times 2{\cdot}0809 = x$$

$$\therefore \quad \underline{x = 145{\cdot}663 \text{ mm.}}$$

Example 2 Two holes are to be drilled in a plate as shown in Fig. 4.11(a). Calculate the dimensions *a* and *b*.

By investigating Fig. 4.11(a), it will be found that the right-angled triangle shown in Fig. 4.11(b) may be used for calculation purposes.

To find *a*:
From Fig. 4.11,

$$a = x + 25 \text{ mm.}$$

From Fig. 4.11(b),

$$\sin 25° 25' = \frac{\text{OPPOSITE}}{\text{HYPOTENUSE}}$$

$$\therefore \quad 0{\cdot}4292 = x/60$$

$$\therefore \quad 60 \times 0{\cdot}4292 = x$$

$$= 25{\cdot}752 \text{ mm}$$

$$\therefore \quad a = 25{\cdot}752 + 25{\cdot}000$$

$$= \underline{50{\cdot}752 \text{ mm.}}$$

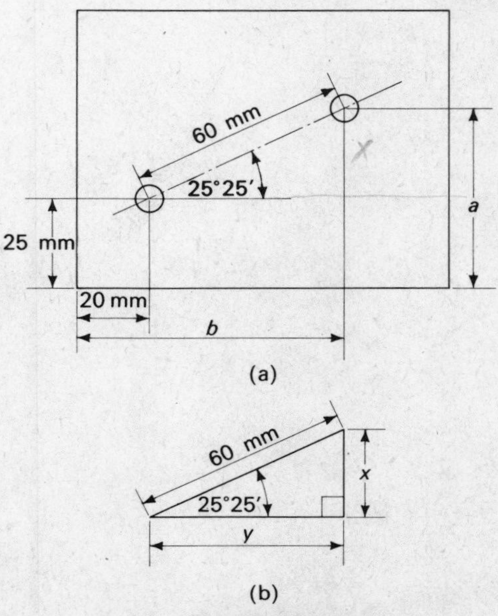

Fig. 4.11 Drilled plate

To find b:
From Fig. 4.11,

$$b = y + 20 \text{ mm.}$$

From Fig. 4.11(b),

$$\cos 25° 25' = \frac{\text{ADJACENT}}{\text{HYPOTENUSE}}$$

$$\therefore \quad 0{\cdot}9032 = y/60$$

$$\therefore \quad 60 \times 0{\cdot}9032 = y$$

$$= 54{\cdot}192 \text{ mm}$$

$$\therefore \quad b = 54{\cdot}192 + 20{\cdot}000$$

$$= \underline{74{\cdot}192 \text{ mm.}}$$

Example 3 A flange has four holes equally spaced around a pitch circle of 80 mm diameter as shown in Fig. 4.12(a). Calculate the centre distance C between adjacent holes.

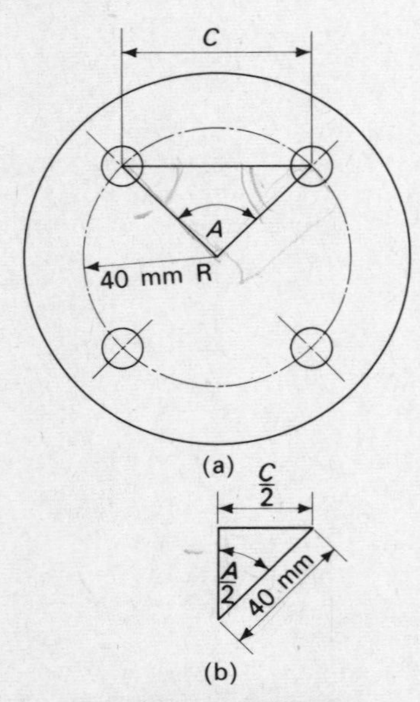

Fig. 4.12 Drilled flange

By investigating Fig. 4.12(a), it will be found that the right-angled triangle shown in Fig. 4.12(b) may be used for calculation purposes.

$$\text{Angle } A = 360°/4 = 90°$$

$$\therefore \quad \text{angle } A/2 = 45°$$

From Fig. 4.12(b),

$$\sin 45° = \frac{\text{OPPOSITE}}{\text{HYPOTENUSE}}$$

$$\therefore \quad 0{\cdot}7071 = (C/2)/40$$

$$\therefore \quad 0{\cdot}7071 = C/80$$

$$\therefore \quad 80 \times 0{\cdot}7071 = C$$

$$\therefore \quad \underline{C = 56{\cdot}568 \text{ mm.}}$$

4.5 Pitch and lead calculations

Many moving parts of measuring instruments and machine tools can be accurately positioned and controlled by the axial movement of a screw or nut. This principle was discussed in section 3.5 and the terms *lead* and *pitch* of a screw thread were defined. The degree of accuracy obtained by this method of controlling movement may be calculated as shown in the following examples.

Example 1 A bench micrometer used in an inspection department is shown in Fig. 4.13.

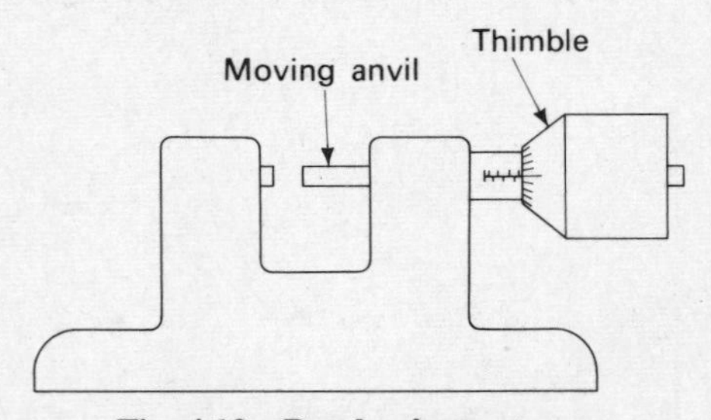

Fig. 4.13 Bench micrometer

The moving anvil is controlled by a single start screw and the thimble is graduated into 100 equal divisions. If the instrument has a measuring accuracy of 0·005 mm, calculate the lead of the screw.

Note that the accuracy of the micrometer will be equal to 1 division on the graduated thimble.

$$\therefore \quad 1 \text{ division on thimble} = 0{\cdot}005 \text{ mm}$$

$$\therefore \quad 1 \text{ complete revolution of thimble} = 100 \times 0{\cdot}005$$

$$= 0{\cdot}5 \text{ mm}$$

$$\therefore \quad \underline{\text{lead of screw} = 0{\cdot}5 \text{ mm.}}$$

Alternatively:

lead of screw (mm) = accuracy (mm) × number of graduations on thimble

$= 0{\cdot}005 \times 100$

$\therefore$ lead of screw = 0·5 mm.

Example 2 The position of the cross-slide of a certain machine tool is controlled by a single start screw, as shown in Fig. 4.14.

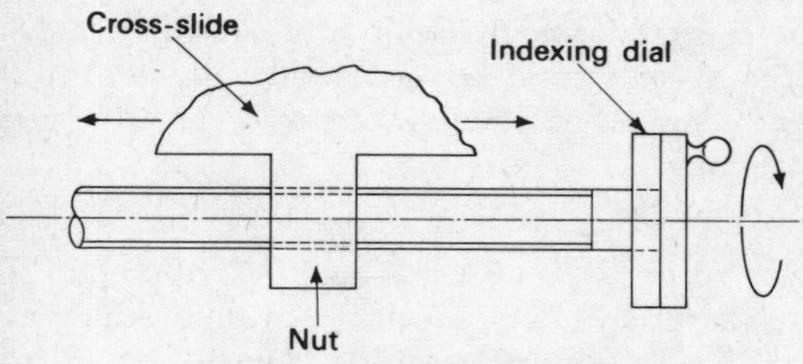

Fig. 4.14 Cross-slide control

If the screw has a pitch of 3 mm, determine:

(a) The lead of the screw
(b) The number of divisions required on the graduated indexing dial to permit control of movement accurate to 0·05 mm.

(a) For single start threads:

lead = pitch

$\therefore$ lead of screw = 3 mm.

(b) Lead of screw (mm) = accuracy (mm) × number of graduations on dial

$3 = 0{\cdot}05 \times$ number of graduations on dial

$\therefore$ 3/0·05 = number of graduations on dial = 60.

Example 3 The table of a certain machine tool is raised by a 2 start elevating screw having a pitch of 6 mm.

Calculate the distance the table is raised when the elevating screw is turned through 8 complete revolutions.

Movement (mm) = lead of screw (mm) × number of revs.

= (number of starts × pitch) × number of revs.

$= 2 \times 6 \times 8$

$\therefore$ movement of table = 96 mm.

4.6 Vernier scales

The principle of the vernier scale was discussed in section 3.6. It was found that the addition of a vernier scale improved the accuracy of any measuring instrument.

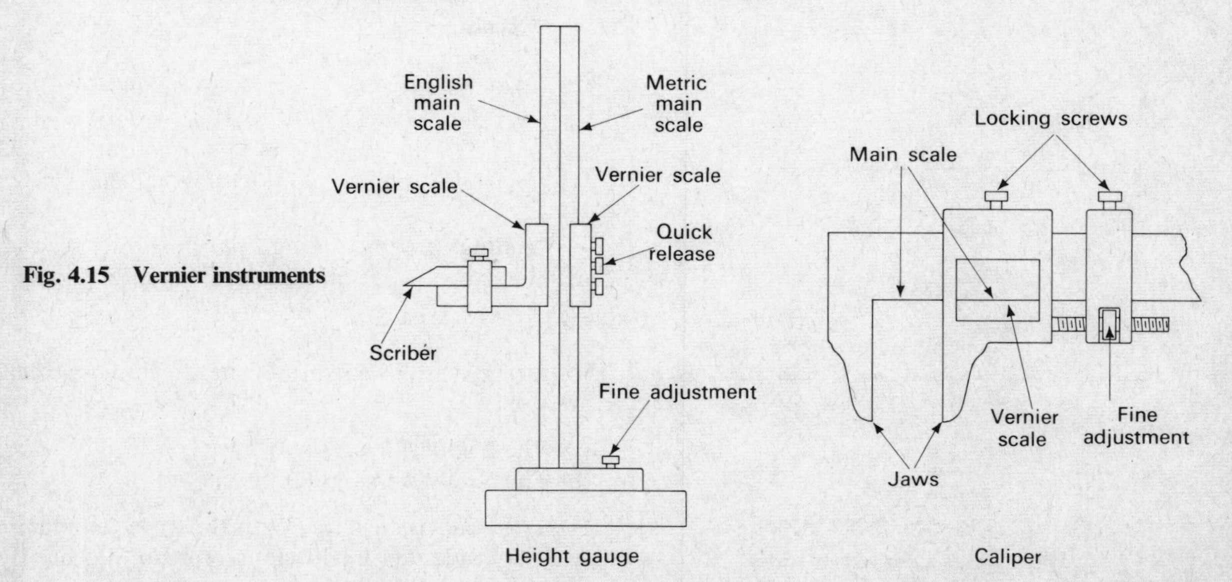

Fig. 4.15 Vernier instruments

Vernier scales may be fitted to micrometers, but are more commonly found on calipers and height gauges. A typical vernier caliper and height gauge are shown in Fig. 4.15.

Both English and metric scales will be found on such instruments, and the different layouts of these scales must be thoroughly understood.

ENGLISH SCALES. Two types of English vernier scale are used:

(a) the 25-division scale
(b) the 50-division scale.

(a) *The 25-division scale.* With this type the main scale is graduated in divisions of $\frac{1}{40}$ inch (0·025 in.). The vernier scale has a total length of 0·6 in. graduated in 25 equal divisions as shown in Fig. 4.16(a).

$$\text{Vernier accuracy (in.)} = \text{1 division on main scale (in.)} - \text{1 division on vernier scale (in.)}$$

$$= 0{\cdot}025 - 0{\cdot}6/25$$

$$= \frac{0{\cdot}625 - 0{\cdot}600}{25}$$

$$= 0{\cdot}025/25 = 0{\cdot}001$$

$\therefore$ <u>vernier accuracy = 0·001 in.</u>

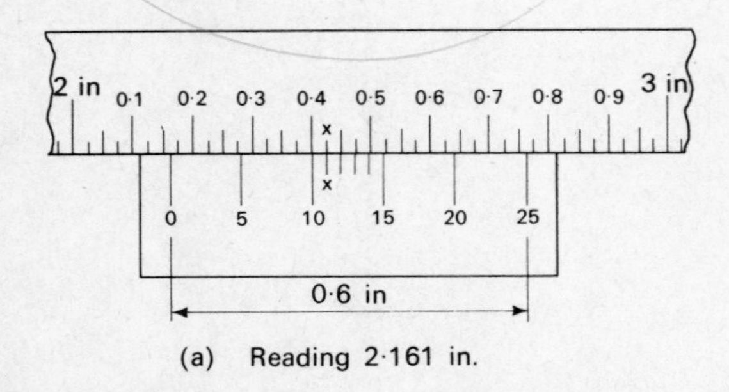

(a) Reading 2·161 in.

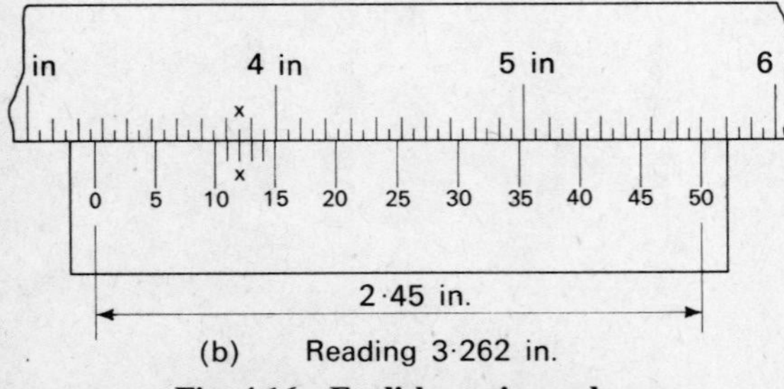

(b) Reading 3·262 in.

Fig. 4.16 English vernier scales

The reading shown is obtained as follows:

		In.
inches:	2	= 2·000
0·1 inch:	1	= 0·100
0·025 inch:	2	= 0·050
vernier div:	11 = 11 × 0·001	= 0·011
	Reading	= 2·161

(b) *The 50-division scale.* With this type the main scale is graduated in divisions of $\frac{1}{20}$ inch (0·050 in.). The vernier scale has a total length of 2·45 in. graduated in 50 equal divisions as shown in Fig. 4.16(b).

Note that:

$$\text{Vernier accuracy (in.)} = \text{1 division on main scale (in.)} - \text{1 division on vernier scale (in.)}$$

$$= 0{\cdot}050 - 2{\cdot}45/50$$

$$= \frac{2{\cdot}500 - 2{\cdot}450}{50}$$

$$= 0{\cdot}050/50 = 0{\cdot}001$$

$\therefore$ <u>vernier accuracy = 0·001 in.</u>

The reading shown is obtained as follows:

		In.
inches:	3	= 3·000
0·1 inch:	2	= 0·200
0·05 inch:	1	= 0·050
vernier div:	12 = 12 × 0·001	= 0·012
	Reading	= 3·262

Readings obtained using English vernier scales may be converted to millimetres by multiplying by 25·4

METRIC SCALES. Two types of metric vernier scale are used:

(a) The 25-division scale.
(b) The 50-division scale.

(a) *The 25-division scale.* With this type the main scale is graduated in divisions of $\frac{1}{2}$ mm (0·5 mm).

The vernier scale has a total length of 12 mm graduated in 25 equal divisions as shown in Fig. 4.17(a).

Note that:

$$\begin{array}{l} \text{Vernier} \\ \text{accuracy} \\ \text{(mm)} \end{array} = \begin{array}{l} \text{1 division on} \\ \text{main scale} \\ \text{(mm)} \end{array} - \begin{array}{l} \text{1 division on} \\ \text{vernier scale} \\ \text{(mm)} \end{array}$$

$$= 0{\cdot}5 - 12/25$$

$$= \frac{12{\cdot}5 - 12}{25}$$

$$= 0{\cdot}50/25 = 0{\cdot}02$$

$\therefore$ vernier accuracy = 0·02 mm.

The reading shown is obtained as follows:

		mm
centimetres:	4	= 40·00
millimetres:	3	= 3·00
0·5 mm:	1	= 0·50
vernier div:	7 = 7 × 0·02	= 0·14
	Reading	= 43·64.

(b) *The 50-division scale.* With this type the main scale is graduated in divisions of 1 mm. The vernier scale has a total length of 49 mm graduated in 50 equal divisions as shown in Fig. 4.17(b).

Note that:

$$\begin{array}{l} \text{Vernier} \\ \text{accuracy} \\ \text{(mm)} \end{array} = \begin{array}{l} \text{1 division on} \\ \text{main scale} \\ \text{(mm)} \end{array} - \begin{array}{l} \text{1 division on} \\ \text{vernier scale} \\ \text{(mm)} \end{array}$$

$$= 1{\cdot}0 - 49/50$$

$$= \frac{50{\cdot}0 - 49{\cdot}0}{50}$$

$$= 1{\cdot}00/50 = 0{\cdot}02$$

$\therefore$ vernier accuracy = 0·02 mm.

The reading is obtained as follows:

		mm
centimetres:	3	= 30·00
millimetres:	1	= 1·00
vernier div:	18 = 18 × 0·02	= 0·36
	Reading	= 31·36.

ANGULAR MEASUREMENT. A vernier protractor may be used for accurate measurement of angles. Such an instrument is shown in Fig. 4.18.

The main scale is graduated in divisions of 1° from 0° to 90° each way. The vernier scale is graduated such that 12 of its divisions span 23° on the main scale as shown in Fig. 4.19.

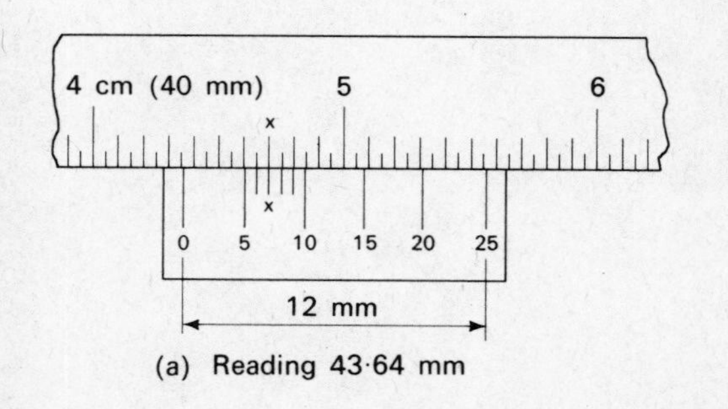

(a) Reading 43·64 mm

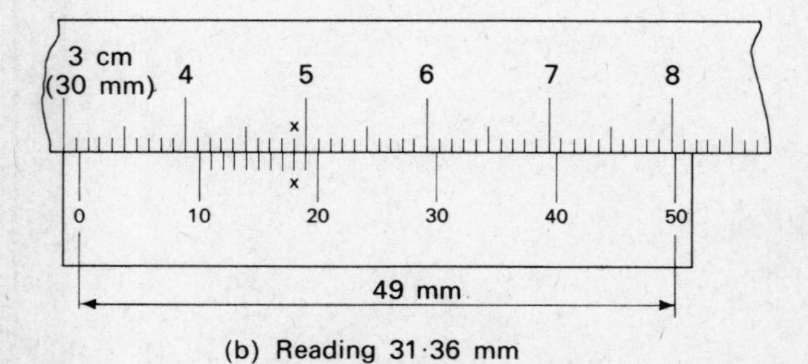

(b) Reading 31·36 mm

Fig. 4.17 Metric vernier scales

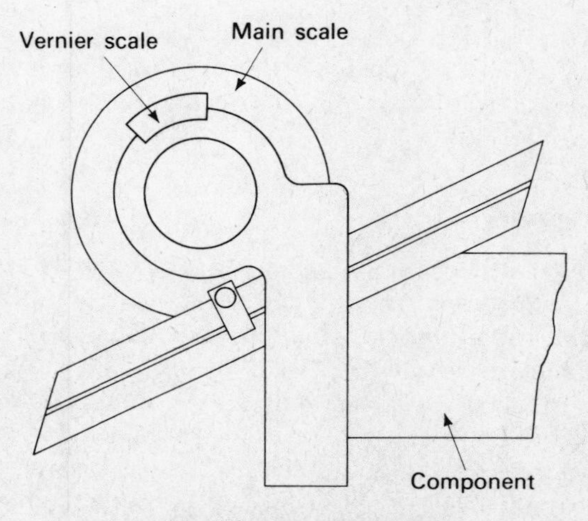

Fig. 4.18 Vernier protractor

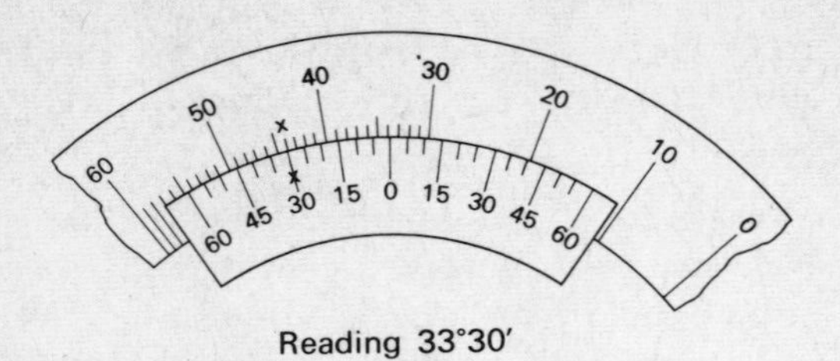

Fig. 4.19 Vernier protractor scales

The accuracy of this instrument must be calculated in a slightly different way from that of all other verniers, because each single vernier division is larger than each single main scale division.

$$\therefore \quad \text{Vernier accuracy } (^\circ) = \text{2 divisions on main scale } (^\circ) - \text{1 division on vernier scale } (^\circ)$$

$$= 2 - 23/12$$

$$= \frac{24 - 23}{12}$$

$$= 1/12^\circ$$

$$\therefore \quad \underline{\text{vernier accuracy} = 5 \text{ minutes.}}$$

The reading shown is obtained as follows:

	degrees	minutes
degrees: 33 =	33	0
minutes: 30 = (on vernier scale)		30
Reading =	33	30

Note that the vernier scale is read in the same direction from the zero as the degrees on the main scale.

Summary

Most materials expand when heated, thus producing increases in length. Such an increase is called linear expansion and may be calculated as explained in section 4.1.

Systems of *limits and fits* indicate suitable limits of size for mating components to provide a required type of fit. The amount of *clearance* or *interference* provided by any type of fit may be calculated as explained in section 4.2.

A sound basic knowledge of *geometry* and *trigonometry* is required for calculations involving angular dimensions on components etc.; these topics are explained in sections 4.3. and 4.4.

Many moving parts of measuring instruments and machine tools can be accurately positioned and controlled by the axial movement of a screw

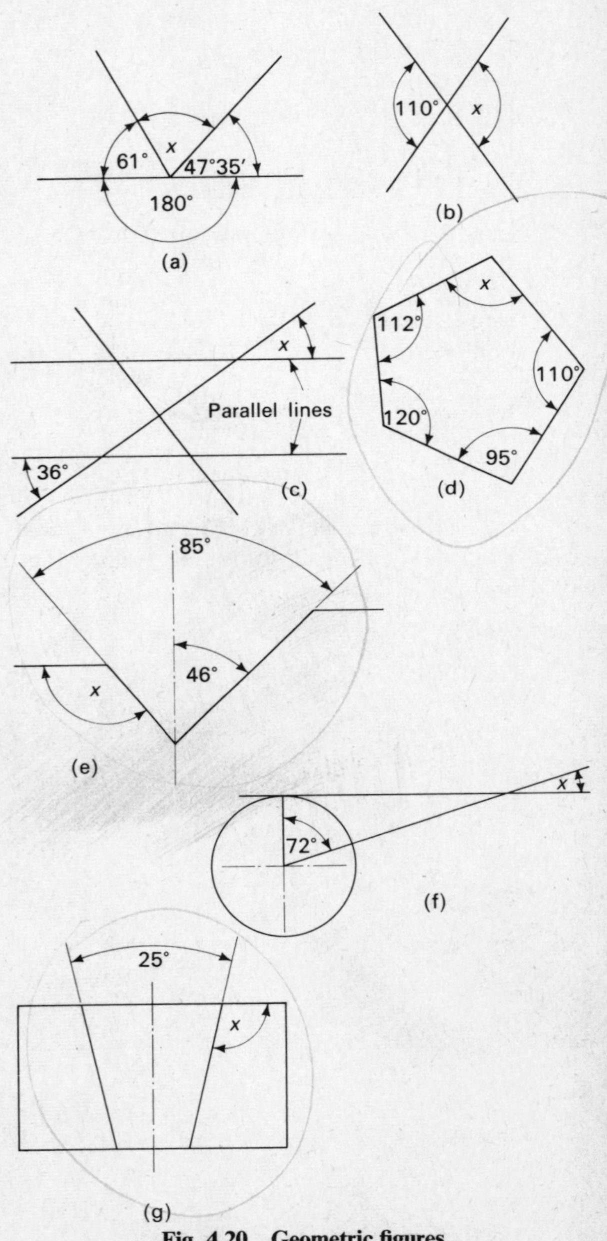

Fig. 4.20 Geometric figures

or nut. Thus the *lead* of a screw can be used to provide accurate control of dimensions.

The layout of different English and metric *vernier scales* used for measuring instruments must be thoroughly understood. Typical vernier scales and their degrees of accuracy are discussed in section 4.6.

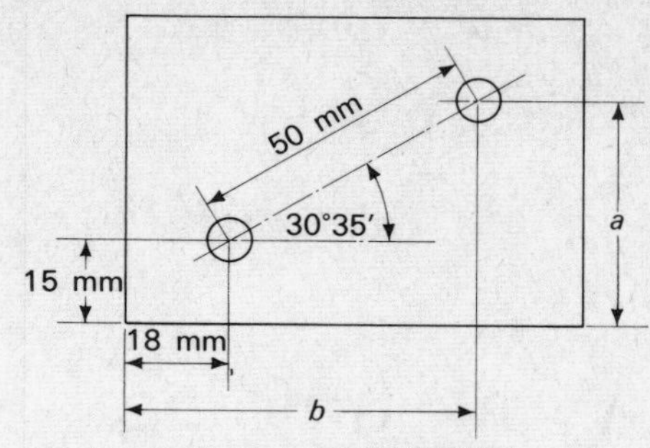

Fig. 4.21 Drilled plate

Questions

1. An end bar used for precise measurement has a length of 600 mm at a standard measuring temperature of 20°C. Calculate its length at a temperature of 30°C given that the material from which the bar is manufactured has a coefficient of linear expansion $\alpha = 1{\cdot}2 \times 10^{-5}/°C$.
2. The diameter of an aluminium piston is measured at a temperature of 20°C and found to be exactly 100 mm. When in service the piston expands to a diameter of 100·96 mm. If for aluminium $\alpha = 2{\cdot}4 \times 10^{5}/°C$, calculate the temperature of the piston when it is in service.
3. The limits of size on a shaft and hole are as follows:

	Hole	Shaft
Maximum dia	30·01 mm	29·98 mm
Minimum dia	30·00 mm	29·97 mm

   Calculate the extremes of fit possible and state the type of fit in each case.
4. A shaft and hole of 60 mm nominal diameter are to be machined to a shaft-basis fit of S7 h6. Referring to Fig. 4.2 determine:
   (a) The type of fit obtained
   (b) The extremes of fit possible.
5. Determine the value of the angle marked *x* in each of the figures shown in Fig. 4.20.
6. Two holes are to be drilled in a plate as shown in Fig. 4.21. Calculate the dimensions *a* and *b*.
7. Six equally spaced holes are drilled around a pitch circle of 60 mm diameter. Calculate the centre distance between adjacent holes.
8. Figure 4.14 shows a simplified diagram of a cross-slide mechanism.
   (a) If the pitch of the single start thread is 5 mm, calculate the number of divisions on the indexing dial needed to provide 0·02 mm increment of movement.
   (b) What form of screw-thread should be used?
   (c) What material should be used to make:
   (i) the screw?
   (ii) the nut?
9. Make neat sketches showing a metric vernier reading of 26·48 mm on:
   (a) A 25-division vernier scale.
   (b) A 50-division vernier scale.
10. Sketch the scales of a vernier protractor reading 15° 50′.

Answers

1. 600·072 mm.
2. 420°C.
3. 0·04 mm clearance, 0·02 mm clearance.
4. (a) Interference (b) 0·072 mm interference, 0·023 mm interference
5. (a) 71° 25′ (b) 110° (c) 36° (d) 103° (e) 129° (f) 18° (g) 102° 30′
6. a = 40·44 mm b = 61·045 mm
7. 30 mm
8. (a) 250.

5. Work holding

Various methods are employed to hold workpieces and components during workshop processes. These methods must offer convenience while ensuring safety and accuracy in use.

This chapter examines the following topics:

5.1 Freedom and restraint

Unless it is restrained (firmly held), any object may be moved in six main paths or directions. These are called the *six degrees of freedom*, and are indicated in Fig. 5.1.

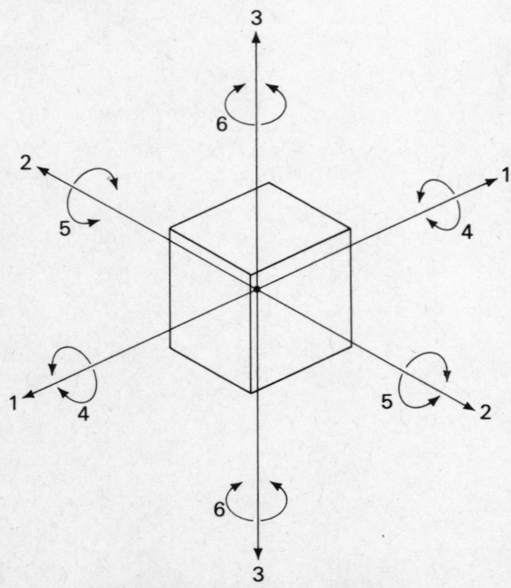

Fig. 5.1 Six degrees of freedom

Notice that the block shown is capable of movement along three straight paths at 90° to each other, and may also rotate about three perpendicular axes.

If these six movements are prevented by applying *restraints*, the block will be unable to move in these or any other directions. Therefore any workpiece or component can be accurately and firmly positioned by *locating* and *clamping* it such that the six degrees of freedom are restrained.

5.2 Principles of location

Suitable *location* is required to position a component or workpiece accurately. This may be provided by various means, including flat datum surfaces, cylindrical plugs and stop devices. The form of location used may be such as to position and secure a component, but often securing is achieved by additional means such as clamps, which are discussed later.

The minimum number of restraints needed to locate a component accurately may be illustrated by returning to the block considered earlier.

By holding the block down on three *locators*, the 3rd, 4th and 5th degrees of freedom may be restrained as shown in Fig. 5.2.

The 1st and 6th degrees of freedom may now be restrained by holding the block against two further locators positioned at one end as shown in Fig. 5.3.

Finally the 2nd degree of freedom may be restrained by holding the block against a sixth and last locator as shown in Fig. 5.4.

Although six separate locators are rarely used for positioning workpieces and components, the actual methods employed must apply the six restraints provided by the location shown in Fig. 5.4.

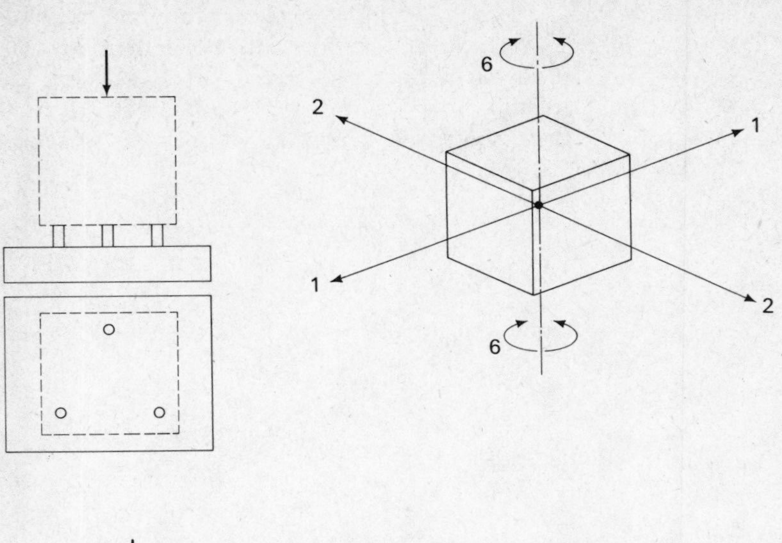

Fig. 5.2 First stage of location

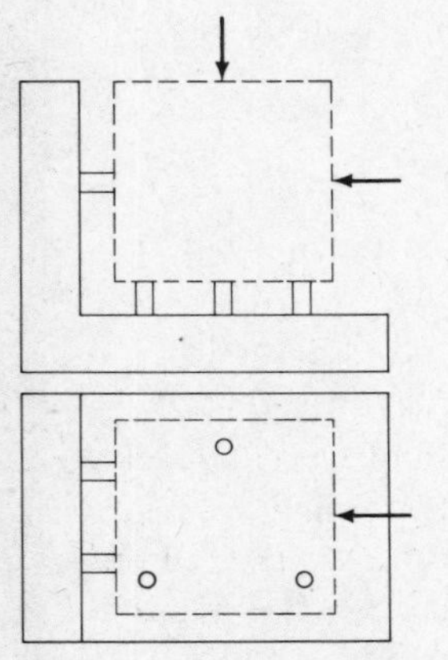

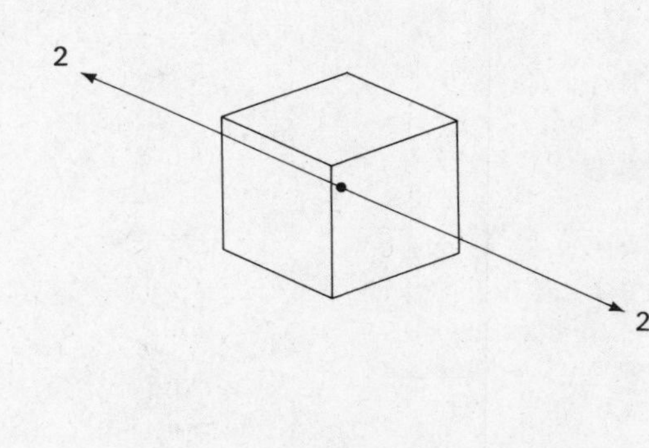

Fig. 5.3 Second stage of location

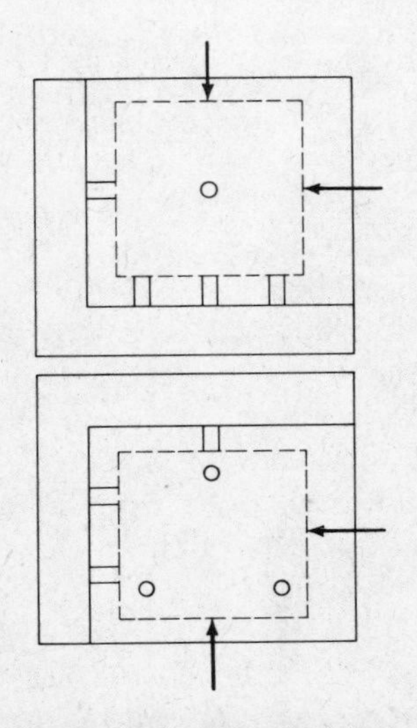

Fig. 5.4 Complete location

5.3 Principles of clamping

Having accurately located a workpiece or component, it will be necessary to hold it securely in this position. This may entail the use of an additional clamping device, which must be strong and easy to operate. Such a device must be positioned safely, e.g., clear of cutters, etc., and should, wherever possible, be arranged such that it applies pressure against a locator and not on an unsupported area. Figure 5.5 illustrates correct and incorrect clamping arrangements.

A clamping arrangement as shown in Fig. 5.5(b) may have two undesirable effects:

(a) The workpiece may become distorted during clamping and a surface machined under such conditions may well be 'bowed' when the clamp is released

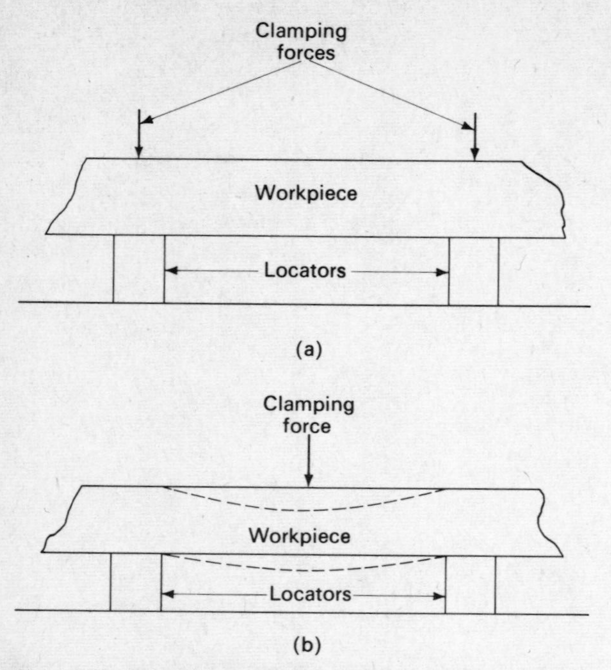

Fig. 5.5 Clamping arrangements: (a) correct; (b) incorrect

(b) 'Springing' can occur when a workpiece is machined under such conditions, and this may well cause vibration, resulting in an unsatisfactory cutting action and poor surface finish.

SUPPLEMENTARY SUPPORTS. When it is not possible to arrange for a clamp to act directly against a locator, an additional or *supplementary support* must be provided, to enable the workpiece to withstand the clamping force. This support must be adjusted barely to contact the workpiece, in order that location is not disturbed. Even when 'ideal' clamping arrangements are possible, supplementary supports may well be required to prevent the workpiece from distorting, due to cutting forces, during a machining process.

5.4 Location and clamping devices

A wide variety of standard devices is used for locating and clamping workpieces and components, the following being common examples:

(a) DOWELS. These are silver steel pins which are accurately ground to a required diameter. Dowels are used to *locate* two parts of a component together, whereupon they may be secured by such means as nuts and bolts. At least two dowels are required to position mating parts accurately, and these should be arranged well apart, as shown in Fig. 5.6.

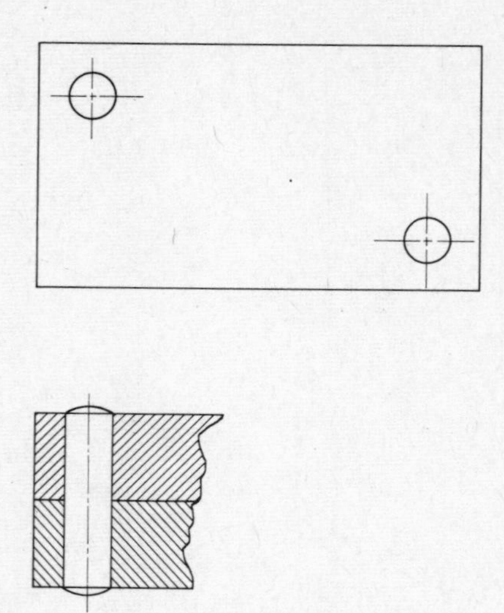

Fig. 5.6 Dowel location

A single dowel used to align the plates shown in Fig. 5.6 would not provide complete location, i.e., rotation about the dowel's axis would still be possible. Thus, a single dowel used in this manner will restrain five degrees of freedom. It follows, therefore, that the two dowels shown in Fig. 5.6 provide more than the six restraints required, i.e., redundant location occurs. For this reason, dowels do not lend themselves to interchangeable manufacture, it being necessary to drill and ream on assembly.

(b) FITTED BOLTS. Screw threads are widely employed for fastening parts together, but are not capable of providing accurate location. Threaded fasteners, such as bolts, are generally intended to pass through clearance holes to ensure ease of assembly. With fitted bolts, however, this is not the case. These are bright bolts, whose shank diameters are ground to be a close fit in the holes which they align.

A fitted bolt is illustrated in Fig. 5.7, showing that it acts as both a dowel and threaded fastener. It is used to *locate* and *clamp* parts together.

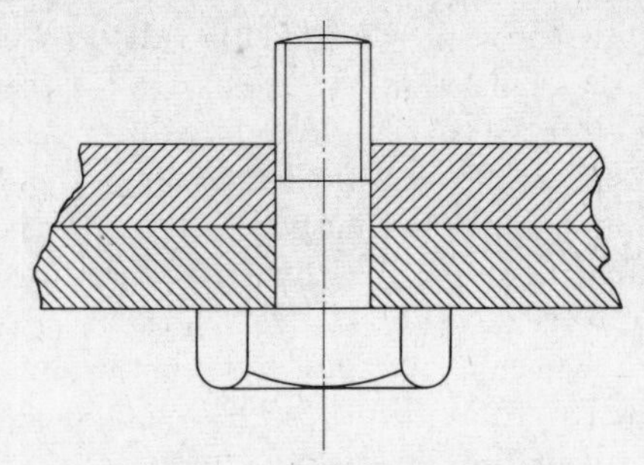

Fig. 5.7 Fitted bolt location

(c) TENONS. A tenon and slot may be used to locate two plane surfaces together, as shown in Fig. 5.8.

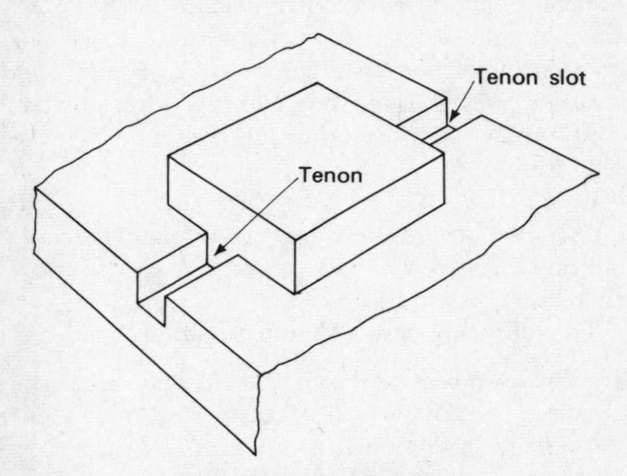

Fig. 5.8 Tenon location

Notice that this arrangement does not provide complete location, endwise movement still being possible.

(d) VICES AND CHUCKS. Basically these are used to clamp workpieces, but they also offer some location. A machine vice, for example, may be secured to a milling machine table in such a way that it will locate a workpiece parallel to the table feed motion. Similarly, a chuck may locate a bar of metal on the centre-line of a lathe. When, however, the endwise position of the bar is of importance, as in capstan and turret lathe processes, additional location is provided by means of a 'bar-stop'. Three basic types of chuck are used for gripping lathe-work, these being three-jaw self centring, four-jaw independent and collet chucks. Of these three types, the three-jaw chuck provides the least satisfactory location, accurate re-setting of work not being possible. If, for example, a bar is to be turned to a certain diameter for its entire length, accurate re-setting can best be achieved by turning between centres. The three-jaw chuck is, however, simple to operate and can accommodate a wide range of workpiece diameters.

The four-jaw chuck is able to provide a more powerful gripping action, due to the independent tightening of each jaw. This independent jaw action also enables workpieces to be located for turning and drilling eccentric features, such as those shown in Fig. 5.9.

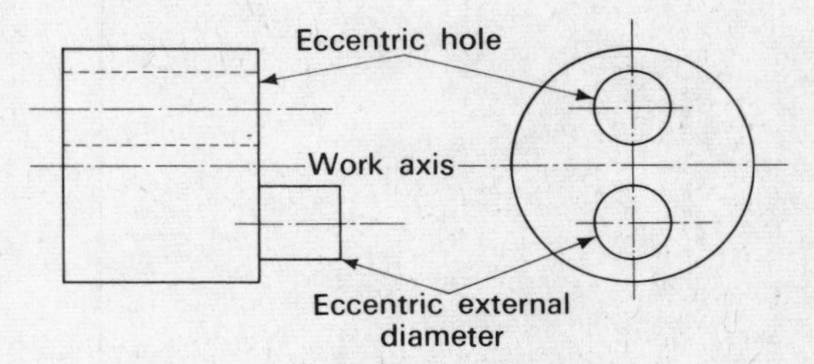

Fig. 5.9 Eccentric features

Various methods may be employed for setting such workpieces and these should be investigated as part of a workshop activity.

The collet chuck is used for gripping bright bars up to about 15 mm diameter. It has an accurate self-centring action and is closed by being pulled into a tapered adaptor as shown in Fig. 5.10.

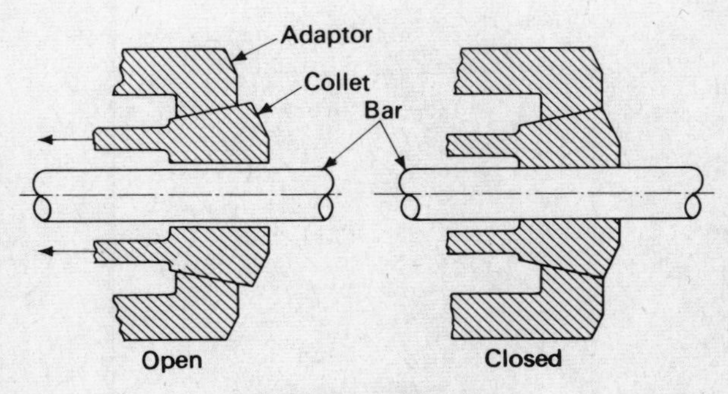

Fig. 5.10 Collet chuck

The collet chuck is simple to operate, provides a firm gripping action, ensures that the work always runs true and does not mark its surface.

(e) MANDRELS. These are used for holding workpieces having finished bores or internal screw threads. Various types of mandrels are available, each locating and holding work between centres in such a way that external diameters may be turned concentric to the bore. Common types of mandrels are shown in Fig. 5.11.

(f) STEADIES. A long, slender workpiece will tend to deflect (bend) during a turning process, due to the cutting forces involved. This may be prevented by using a *supplementary support*, in the form of a *steady*. *Fixed* and *travelling* types of steady are available, these being illustrated in Fig. 5.12.

The *fixed steady* is clamped to the lathe bed at a convenient position, and the three bearing fingers are adjusted to contact the workpiece surface. When, however, a diameter must be turned for the complete length of a workpiece, it will be necessary to reposition the steady during the cut to allow passage of the tool. The *travelling* steady is clamped to the lathe saddle, and is adjusted to contact the workpiece immediately behind the tool.

Fixed steadies are often employed for supporting the end of a long bar while it is faced or drilled. A travelling steady is used to support a bar while a diameter is turned for a considerable length. This steady moves with the tool, and therefore provides continuous support throughout the turning operation.

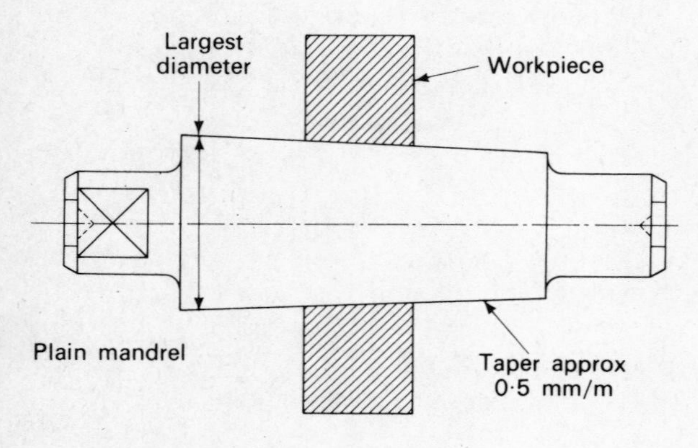

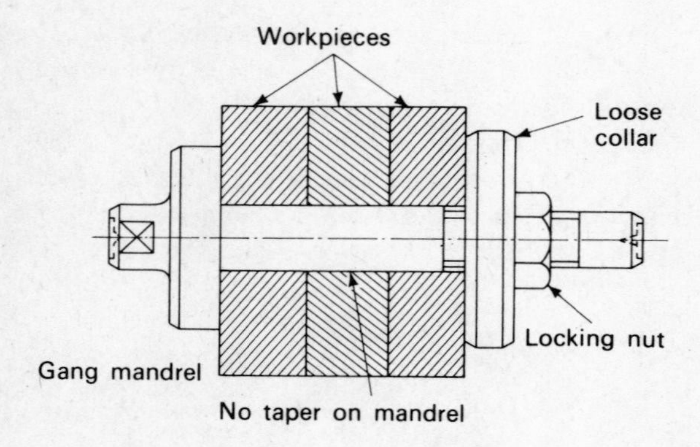

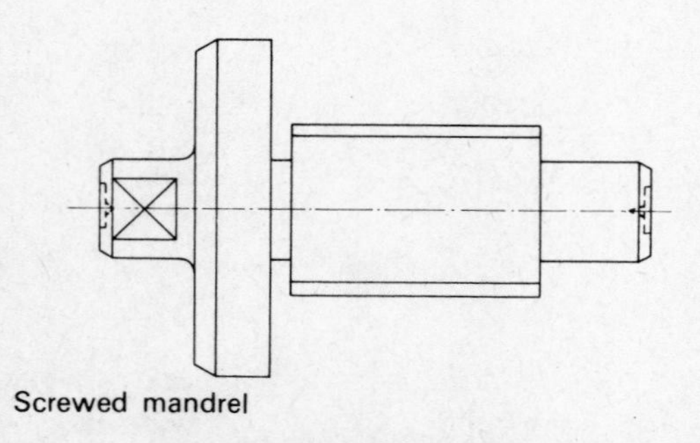

Fig. 5.11 Types of mandrel

(g) STRAP CLAMPS. These are used in many machining processes to *clamp* workpieces. A simple strap clamp is shown in Fig. 5.13.

The following points should be noted:

(i) The *tee bolt* is positioned as close to the workpiece as possible. The reason for this is explained in section 6.1.
(ii) the *toe* and *heel* of the clamp are radiused. This is to ensure that the workpiece will be securely held if the clamp is not parallel to the machine table.

(h) EDGE CLAMPS. These may be used for *clamping* during milling, shaping, and planing processes, and enable the whole top surface of the workpiece to be machined. An edge clamp is shown in Fig. 5.14.

5.5 Examples of location and clamping

The following examples illustrate various methods of locating and clamping:

(a) Work-holding equipment
(b) Tool-holding equipment
(c) Workpieces.

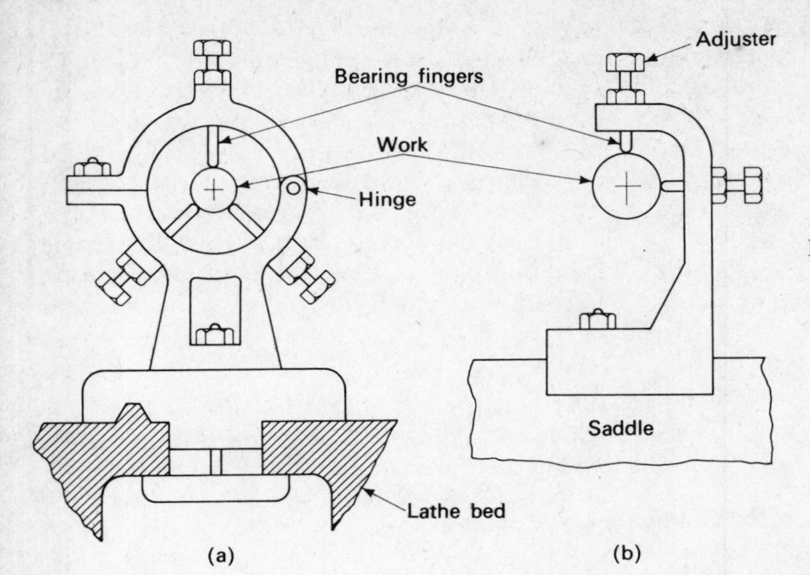

Fig. 5.12 Types of steady: (a) fixed; (b) travelling

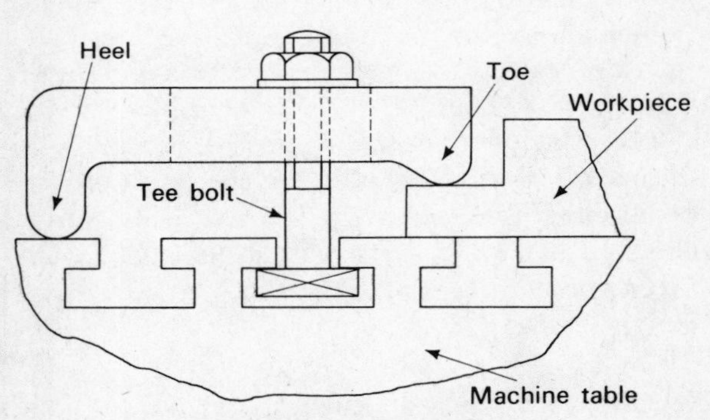

Fig. 5.13 Strap clamp

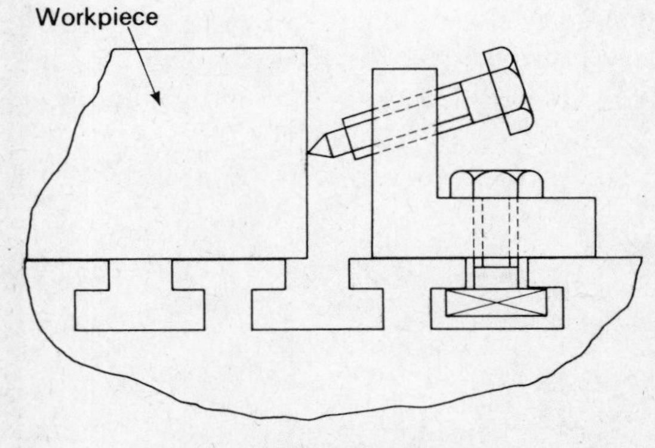

Fig. 5.14 Edge clamp

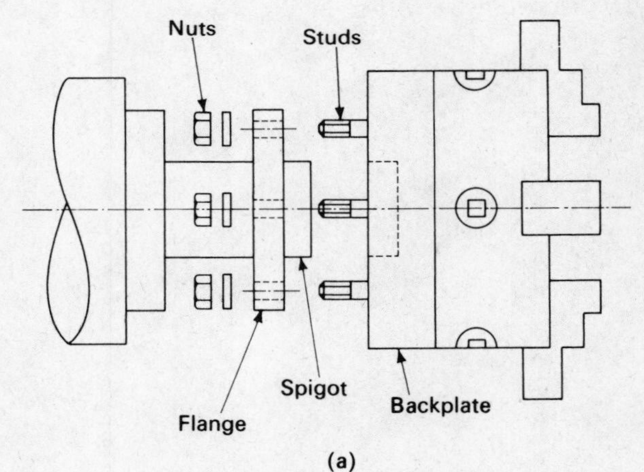

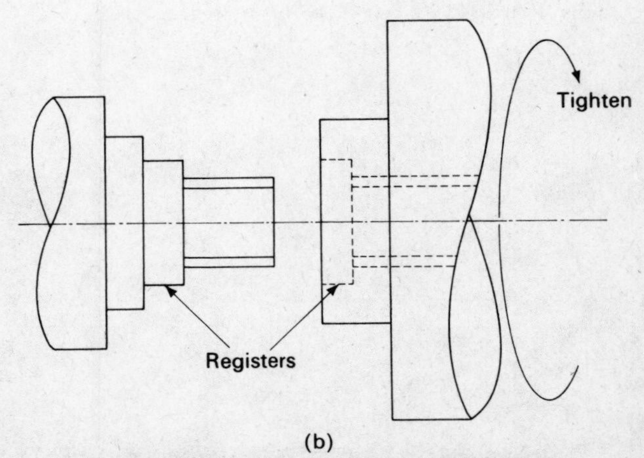

Fig. 5.15 Location and securing of lathe chucks

(a) LOCATION AND CLAMPING OF WORK-HOLDING EQUIPMENT. Two methods of securing chucks to lathe spindles are shown in Fig. 5.15.

With the method shown in Fig. 5.15(a), the chuck is fastened to a flange on the spindle nose by means of studs and nuts. Accurate location is provided by a ground diameter on the spindle nose called a *spigot* or *register*, which mates with a bored recess in the backplate of the chuck. It should be noted that the register, studs and nuts together provide the six necessary restraints discussed earlier.

With the method shown in Fig. 5.15(b) the chuck is screwed directly onto the lathe spindle, accurate location again being provided by means of registers. During a turning operation, the cutting force will tend to tighten the chuck still further onto the spindle nose. If, however, it is required to turn a workpiece with the chuck rotating in a reverse direction, the tendency now will be for the chuck to become loosened and detached from the spindle. This possible danger does not exist when chucks or face plates are secured by studs and nuts or by a third method as shown in Fig. 5.16.

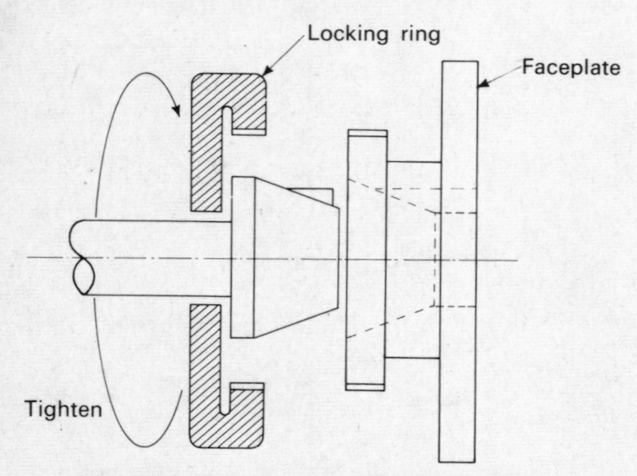

Fig. 5.16 Location and securing of faceplate

With this method the chuck or faceplate is secured by a threaded locking ring and location is provided by mating tapers. Note that with this arrangement the drive is transmitted to the chuck by means of a key.

Machine vices and dividing heads are frequently employed for holding workpieces during milling and shaping processes. Such work-holding equipment is clamped to the machine table by means of tee-bolts and may be located by tenons as shown in Fig. 5.17.

(b) LOCATION AND CLAMPING OF TOOL-HOLDING EQUIPMENT. The spindle and arbor of a horizontal milling machine are illustrated in Fig. 5.18. The arbor is a tool-holding device, its purpose being to carry milling cutters.

Notice that the arbor is located in the spindle nose by mating tapers and a draw-bolt is used to tighten and secure the assembly. This arrangement provides extremely accurate location combined with efficient clamping.

The principle of obtaining accurate location by means of mating tapers is widely employed for positioning tool-holding devices. Self-centring chucks, used for holding twist drills, and holders for end mills are typical examples of devices located in this manner.

(c) LOCATION AND CLAMPING OF WORKPIECES. Various means of providing satisfactory location and clamping were described in section 5.4. Many other devices are available for these purposes, but those previously mentioned are suitable for holding the majority of workpieces, providing they are

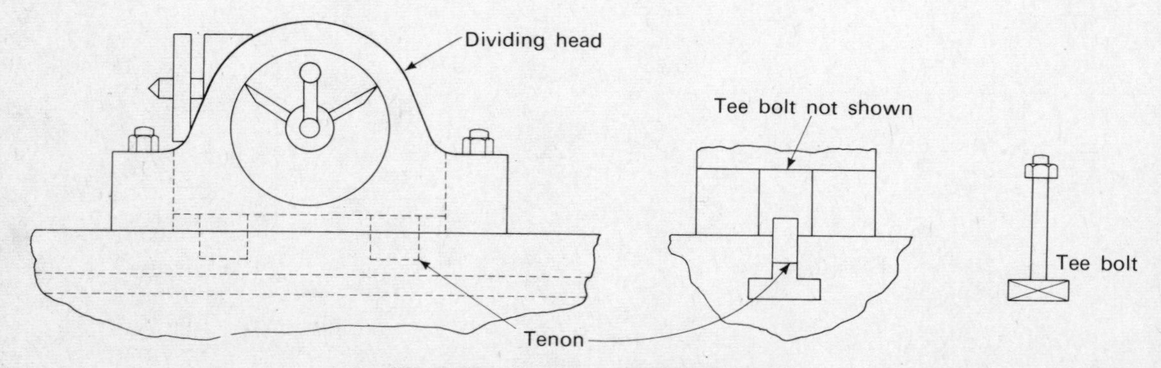

Fig. 5.17 Use of tee bolts and tenons

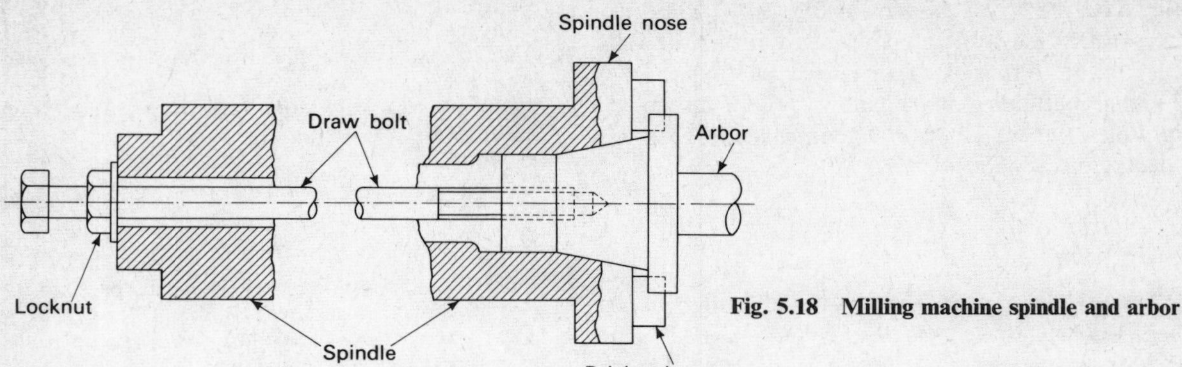

Fig. 5.18 Milling machine spindle and arbor

used correctly. The machine vice, for example, is only intended to grip parallel surfaces on workpieces. It should never be used for gripping on the diameter of a round bar, this clamping arrangement being insecure. Satisfactory methods of holding cylindrical workpieces for drilling and shaping are shown in Figs. 5.19 and 5.20.

Pairs of vee blocks are frequently used for location and clamping on workpiece diameters, and the following points should be noted:

(i) The vee blocks shown in Fig. 5.19 locate the workpiece parallel to the drilling-machine table. This enables the hole to be drilled square to the axis of the workpiece.

(ii) Each clamp and vee block shown in Fig. 5.19 together contact the circumference of the workpiece at three points. The workpiece is, therefore, secured by a similar action to that of a three-jaw chuck.

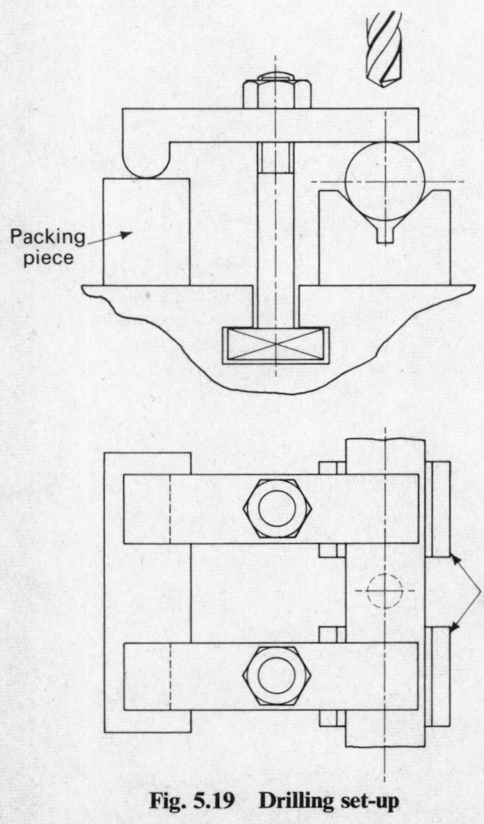

Fig. 5.19 Drilling set-up

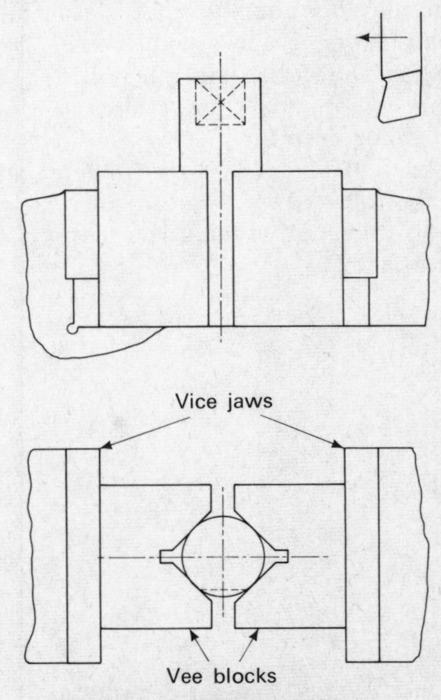

Fig. 5.20 Shaping set-up

(iii) The vee blocks shown in Fig. 5.20 locate the workpiece vertically and secure it by a similar action to that of a four-jaw chuck.

Further methods of work holding will be discussed in Vol. 2 where additional requirements such as 'balancing' are considered.

Summary

When a workpiece or component is required to be held securely, it must be *located* and *clamped* such that its *six degrees of freedom* are eliminated. These six degrees of freedom are paths or directions in which an object may move, unless it is restrained.

Workpieces and components may be suitably located by such means as *dowels, fitted bolts, tenons, registers, mating tapers*, and *vee blocks.* Standard equipment such as *chucks* and *machine vices* may be used both to locate and clamp a wide variety of workpieces, but when these are not suitable, other means of clamping must be employed. *Strap clamps* may be used to secure workpieces during drilling, shaping, milling and planing operations. Such clamps must be positioned to act on a supported area of the workpiece to prevent distortion and vibration from occurring. Supplementary supports may be used to enable a workpiece to withstand clamping and cutting forces. *Fixed* or *travelling steadies*, for example, can provide supplementary support for long slender workpieces during turning operations.

Any method used for work holding must be absolutely safe. Always remember that safety carries no tolerance, nothing less than perfection being satisfactory.

Questions

1. What are the main requirements of a work-holding device?
2. Sketch a typical application of each of the following location devices:
   (a) Dowels
   (b) Fitted bolts
   (c) Tenons
   (d) Registers
   (e) Tapers.
3. When should supplementary supports be used in work-holding arrangements?
4. Sketch a screw jack suitable for providing supplementary support during a milling operation.
5. Compare the uses of fixed and travelling steadies.
6. A hole 40·00/40·05 mm diameter and 80 mm deep is to be produced in the end of a mild steel bar, 800 mm in length, which has been finish turned to 70 mm diameter. Describe how the bar could be set up in the lathe, assuming that it cannot be fed through the headstock spindle.
7. Describe, with the aid of sketches, how the 20 mm diameter could be turned on the workpiece shown in Fig. 5.21.

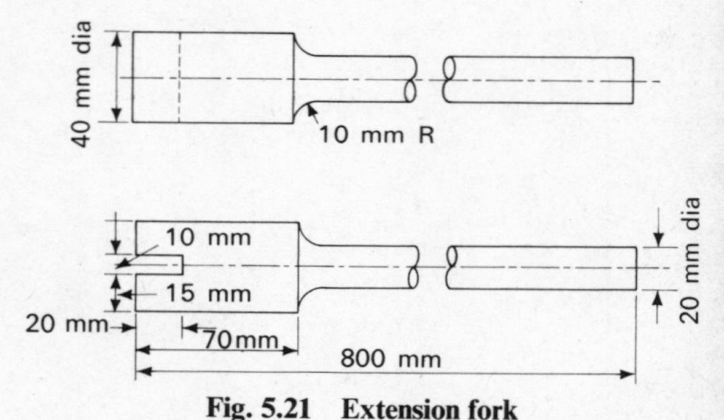

Fig. 5.21 Extension fork

8. Sketch a suitable set-up for milling the slot in the workpiece shown in Fig. 5.21, assuming that the turning operation has previously been completed.

6. Studies associated with work holding

This chapter examines the following topics associated with work holding.

5.1 Forces and moments
5.2 The use and effects of friction.

6.1 Forces and moments

During a machining process a workpiece may tend to move due to the cutting forces acting on it. Clamping forces must, therefore, be provided to prevent this from happening.

The effect of a force may be magnified by *leverage.* A spanner, for example, will provide leverage when tightening a nut. The force applied at the end of the spanner need not be great whilst the nut is being firmly secured. This is because a *turning moment* or *moment of force* is applied to the nut, causing it to rotate about its centre or axis.

The value of this moment of force may be calculated from the formula:

$$\text{Moment of force} = \text{force} \times \text{perpendicular distance,}$$

where the *perpendicular distance* is the shortest distance between the line of action of the force and the axis about which it causes rotation.

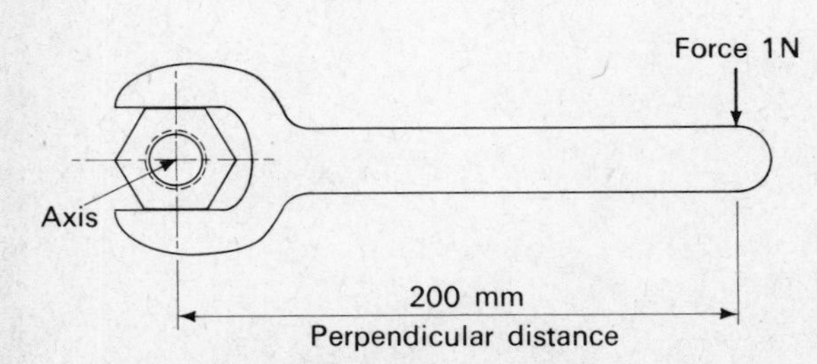

Fig. 6.1 Moment of force

Example Calculate the moment of force applied to the nut shown in Fig. 6.1.

$$\begin{aligned}\text{Moment of force} &= \text{force} \times \text{perpendicular distance}\\ &= 1\ \text{N} \times 200\ \text{mm}\\ &= \underline{200\ \text{N mm.}}\end{aligned}$$

Standard work-holding devices, such as machine vices and lathe chucks, are tightened by applying a force at the end of a handle or chuck key. This produces a moment of force which enables efficient tightening to be carried out with ease.

Example Calculate the moment of force applied to the vice screw by the handle shown in Fig. 6.2.

$$\begin{aligned}\text{Moment of force} &= \text{force} \times \text{perpendicular distance}\\ &= 40\ \text{N} \times 200\ \text{mm}\\ &= \underline{8000\ \text{N mm}} \text{ or } \underline{8\ \text{N m}}\end{aligned}$$

Example Calculate the moment of force, in N m, provided by the chuck key shown in Fig. 6.3.

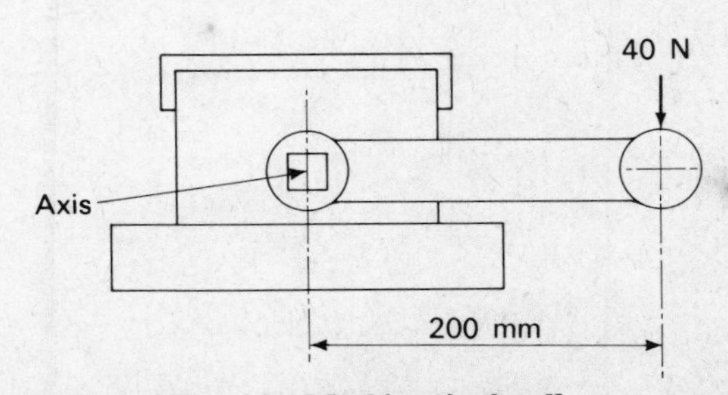

Fig. 6.2 Machine vice handle

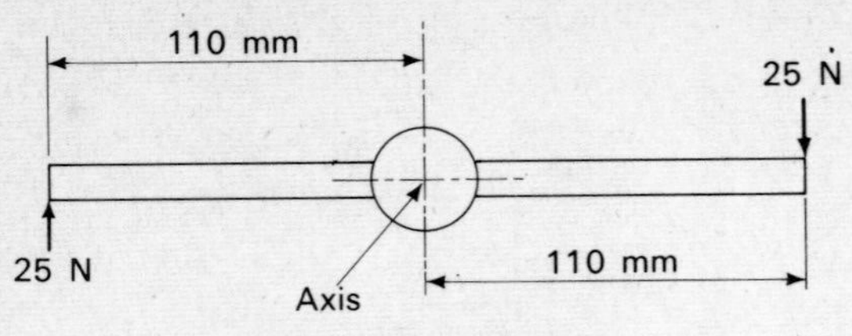

Fig. 6.3 Chuck key

Notice that in this example two forces cause rotation, each producing a moment of force. These assist one another in tightening the chuck and their values must therefore be added together.

$$\therefore \text{Total moment of force} = (25 \times 110) + (25 \times 110)$$
$$= 2750 \text{ N mm} + 2750 \text{ N mm}$$
$$= 5500 \text{ N mm}$$
$$= \underline{5{\cdot}5 \text{ N m.}}$$

It should now be apparent that when leverage is used, the greater the perpendicular distance between the force and axis of rotation, the more effective will be the moment of force produced. This principle is applied to all clamping devices such as the strap clamp shown in Fig. 6.4.

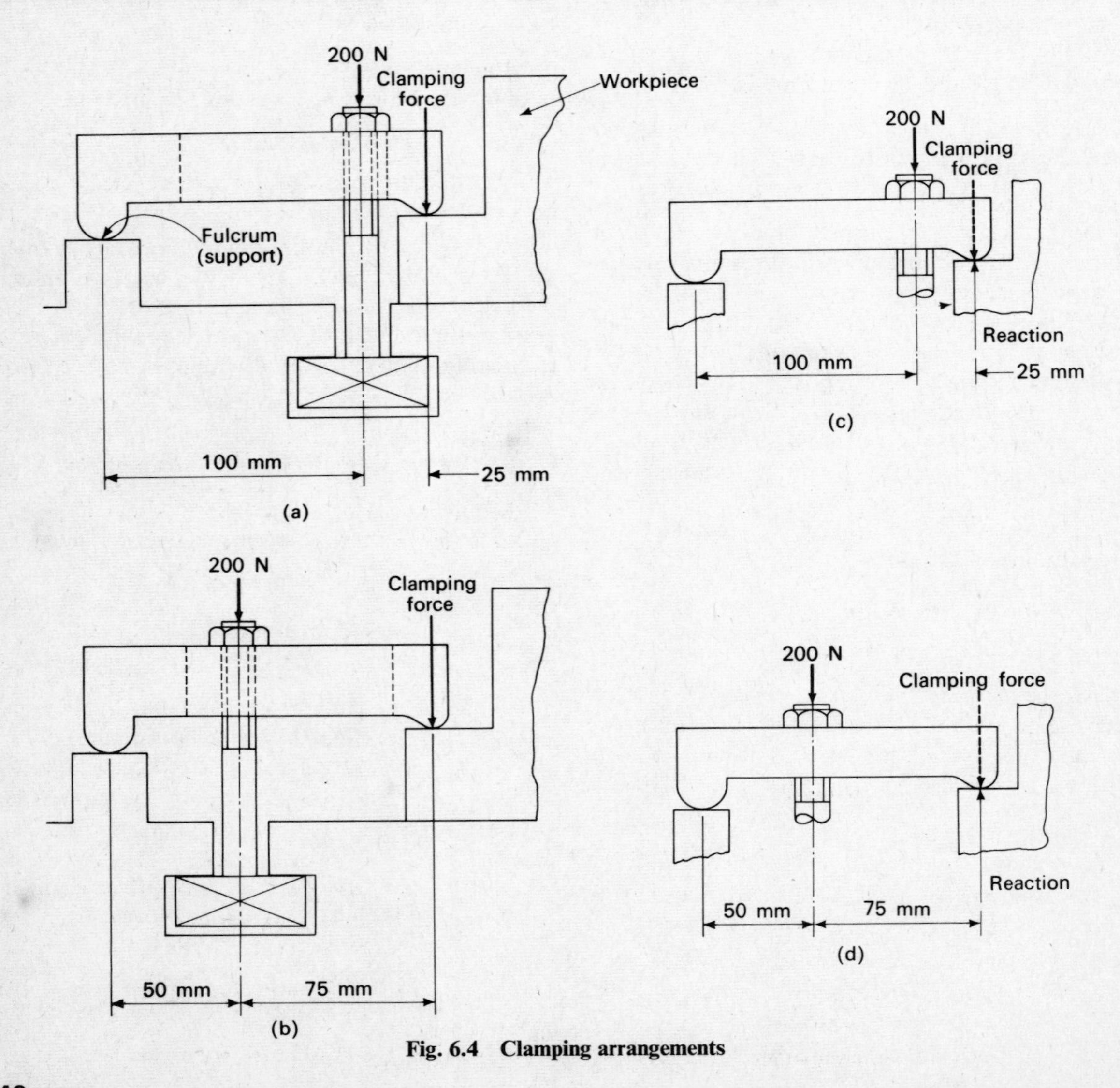

Fig. 6.4 Clamping arrangements

Referring to Fig. 6.4(a)

Moment of force = force × perpendicular distance

$= 200\text{ N} \times 100\text{ mm}$

$= 20\,000\text{ N mm}$

$= \underline{20\text{ N m}}.$

However, with the tee bolt positioned as shown in Fig. 6.4(b):

Moment of force $= 200\text{ N} \times 50\text{ mm}$

$= 10\,000\text{ N mm}$

$= \underline{10\text{ N m}}.$

It should therefore be noted that the closer the tee bolt is positioned to the workpiece, the greater will be the moment of force provided by the clamp. Furthermore, the greater this moment of force, the greater will be the clamping force acting on the workpiece.

Figure 6.4(c) shows that the clamping force acting downwards on the workpiece is resisted by an equal *reaction* acting upward. This reaction tends to rotate the clamp anticlockwise about its fulcrum, whilst the force in the bolt tends to cause clockwise rotation. The value of this reaction may be calculated using the *Principle of Moments.* This states that for a body to be at rest, the sum of the clockwise moments must be equal to the sum of the anticlockwise moments.

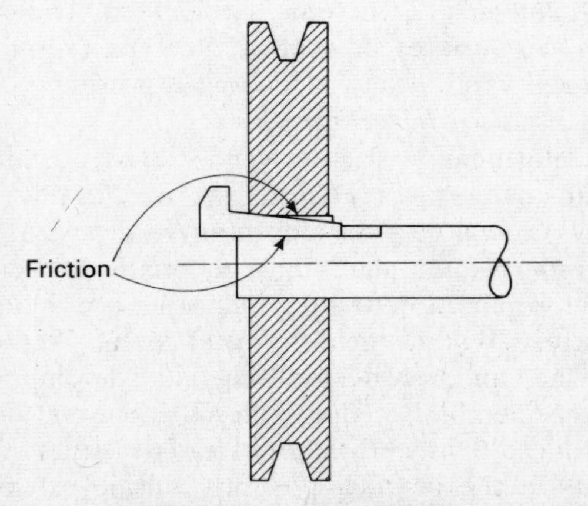

Fig. 6.5 Key held by friction

Therefore:

Sum of clockwise moments = sum of anticlockwise moments

$$200 \times 100 = \text{reaction} \times 125$$

$$\frac{200 \times 100}{125} = \text{reaction} = 160\text{ N}$$

$$\therefore \quad \underline{\text{clamping force} = 160\text{ N}}.$$

This calculation can be repeated for the clamping arrangement shown in Fig. 6.4(d):

Sum of clockwise moments = sum of anticlockwise moments

$$200 \times 50 = \text{reaction} \times 125$$

$$\frac{200 \times 50}{125} = \text{reaction} = 80\text{ N}$$

$$\therefore \quad \underline{\text{clamping force} = 80\text{ N}}.$$

These calculations will emphasize the necessity for positioning the tee bolt of a clamp as close to the workpiece as possible in order to provide the maximum clamping force.

6.2 The use and effects of friction

A rotating bicycle wheel may be brought to a halt by pressing a brake block against the surface of the wheel rim. The wheel's movement is halted due to a force between the surfaces of the brake block and rim. This force is called *friction*, and may be defined as 'a force which tends to resist sliding motion between two surfaces in contact.'

The value of a friction force will depend upon two factors:

(a) The nature of the materials in contact
(b) The total force causing contact between the surfaces.

The materials used for brake blocks and wheel rims, i.e., rubber and steel, cause a high friction force to occur between their moving surfaces. The leverage provided by the brake mechanism increases the pressure between these surfaces and hence an extremely high friction force is produced.

Many devices used for securing or clamping workpieces and components rely on friction for their action. A tapered key, for example, may be used to secure a pulley to a machine spindle as shown in Fig. 6.5.

When driven into position, the key acts as a wedge, which both secures the pulley to the spindle and transmits a drive between them. The key is prevented from working loose by the presence of friction at its surface.

In a similar manner threaded fasteners rely on friction to keep them securely in position. A nut and bolt are shown in Fig. 6.6 to illustrate the presence of friction.

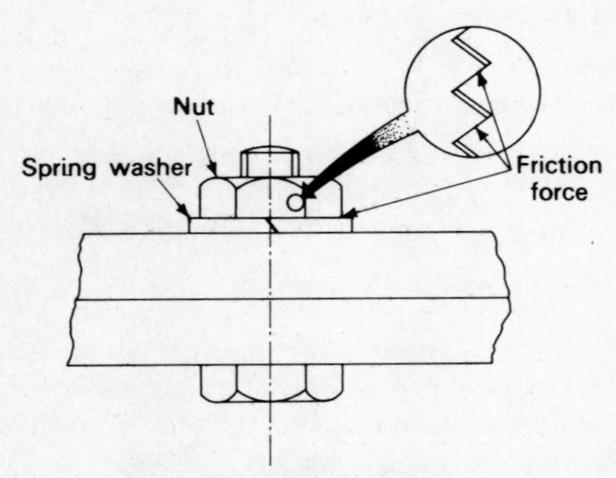

Fig. 6.6 Friction in threaded fasteners

Notice that when the nut is tightened, pressure acts on one side of the thread flanks. This results in friction between the nut and bolt threads which tends to prevent the nut from working loose.

By positioning a spring washer beneath the nut it can be ensured that the pressure acting on the thread flanks is maintained, even when the nut and bolt are subjected to vibration. Referring to Fig. 6.6 it should be noted that friction also occurs at the face of the nut, further preventing it from becoming loosened.

Considerable friction can exist between metal surfaces in contact, this force being both useful and necessary when securing workpieces for machining operations etc. A lathe chuck, for example, is able to grip a workpiece and prevent it from slipping during a turning operation. This is only possible due to the presence of friction between the contacting surfaces of the chuck jaws and workpiece.

A clamping arrangement on a centre-lathe faceplate is shown in Fig. 6.7. This set-up relies on frictional forces to keep the workpiece in position when a hole is bored in it.

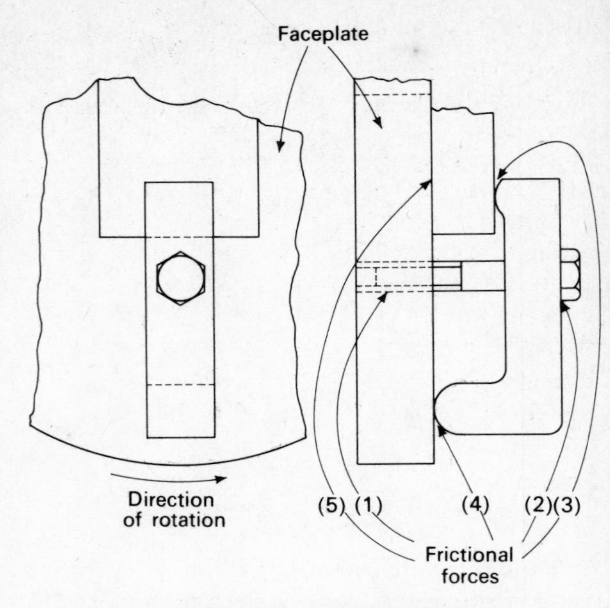

Fig. 6.7 Friction in a clamping arrangement

The frictional forces indicated in Fig. 6.7 are:

(1) Between the thread flanks
(2) At the face of the bolt head (these prevent the clamp bolt from becoming loose)
(3) Between the clamp toe and workpiece
(4) Between the clamp heel and faceplate
(5) Between the workpiece and faceplate (these prevent the clamp and the workpiece from moving).

The force of friction is used to advantage in many securing and clamping arrangements. It is also employed as a driving force in devices such as belts and pulleys, where any slipping action between the two surfaces in contact must be prevented. Typical applications of friction as a driving force are discussed in section 8.1.

Unfortunately, friction cannot always be used to advantage by the engineer. At times its presence may, in fact, be a distinct disadvantage. Bearings carrying rotating shafts must be suitably lubricated to reduce friction to a minimum. Lack of lubrication would allow metal to metal contact, between the shaft and bearing, resulting in the development of considerable heat due to the high frictional forces produced. This, in turn, could lead to seizure of the shaft in the bearing, possibly causing extensive damage.

Similarly, many supporting devices used in work-holding set-ups must be protected from the harmful effects of friction. A dead centre held in a lathe tail-stock will be subjected to extremely high frictional forces when supporting a bar during a turning process. Suitable lubricating grease must therefore be employed to reduce friction between the surfaces in contact. During a lengthy turning operation the tailstock pressure may also require adjustment to allow for expansion of the work, as explained in section 3.2. Similar precautions must be taken to reduce friction to a minimum when steadies are used to support workpieces.

Summary

The effect of a force may be magnified by leverage. The turning effect produced when a force is applied at the end of a lever, such as a spanner, is called a *moment of force* and may be calculated as explained in section 6.1.

Friction may be considered as a force which tends to resist sliding motion between two surfaces in contact. The value of a frictional force will depend upon the nature of the materials in contact and the pressure holding them together. Many devices used for securing or clamping workpieces and components, rely on friction for their action. The ability of friction to resist movement is, for example, particularly useful in screw threads and clamping arrangements used for work holding.

Questions

1. What is a *turning moment* or *moment of force*?
2. A *C* spanner is used to tighten the locking ring of a lathe chuck as shown in Fig. 6.8.

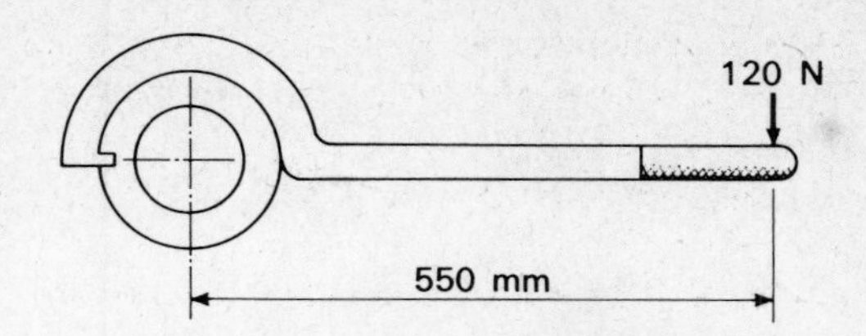

Fig. 6.8 *C* spanner

Calculate the moment of force in Nm, applied to the locking ring.

3. Calculate the clamping force, in N, acting on the workpiece shown in Fig. 6.9, and indicate two unsatisfactory features of this clamping arrangement.

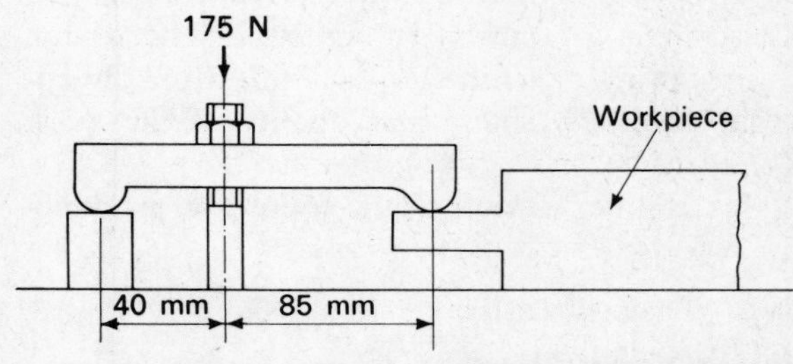

Fig. 6.9 Unsatisfactory clamping arrangement

4. A moment of force of 15 N m is used to tighten a lathe chuck. If the handle of the chuck key has an overall length of 300 mm, calculate the total force in N applied by the operator.
5. What is a frictional force?
6. Give three examples where friction is of advantage to the engineer in work-holding arrangements.
7. Discuss a possible disadvantage of friction in work-piece-supporting devices.

Answers

2. 66 N m **3.** 56 N **4.** 50 N

7. Machine tools

A machine tool is a power-operated device designed for the production of workpieces, by such means as the removal of unwanted material. The design features of the machine tool will affect the dimensional accuracy, shape and quality of the work produced.

This chapter examines the following machine-tool features and considerations:

7.1 Basic structure
7.2 Alignment
7.3 Relative movement between cutting tool and workpiece.

7.1 Basic structure of machine tools

A machine tool must have the following:

(a) A means of holding the tool, e.g., toolpost, arbor, etc.
(b) A means of holding the workpiece, e.g., chuck, vice, etc.
(c) A source of power, e.g., a mains-supplied motor
(d) A system for transmitting power to the tool and work-holding devices
(e) A means of controlling the relative motion between the tool and workpiece
(f) A rigid frame.

Various tool-holding and work-holding devices have been discussed in chapter 5. Therefore at this stage consideration will be given to the remaining machine-tool requirements listed.

All machine tools consist of several parts supported and located on a main frame. This frame may be called a *bed*, as in the case of lathes and grinding machines, or a *body* when referring to shapers and milling machines. In all cases the design of the machine frame is of critical importance, as it must satisfy two major objectives:

(a) It must permit accurate location of all other parts of the machine in relation to each other, to ensure the production of satisfactory workpieces
(b) It must absorb vibration set up when the machine is running, to prevent poor surface finish on workpieces.

RIGIDITY. To ensure accurate location of machine-tool parts with relation to each other, the frame must be rigid. This is to prevent distortion due to cutting forces and parts of the machine in motion. To provide rigidity, machine-tool frames are usually either of *ribbed* or *box* design. Typical sections are shown in Fig. 7.1 to illustrate these terms.

A machine-tool frame may distort either by bending or twisting. For the same amount of material, a box section is approximately twelve times more rigid than a ribbed section when subjected to twisting or torsion, and four times more rigid against bending.

Consider the forces carried by a lathe bed during a turning operation. These are:

(a) Downward pressure of the work acting on the tool
(b) Outward pressure forcing the tool away from the work in a horizontal direction.

These are shown in Fig. 7.2.

The net result of these forces is largely one of torsion, and the bed is designed to resist twisting. Although the box design is best suited to this, a

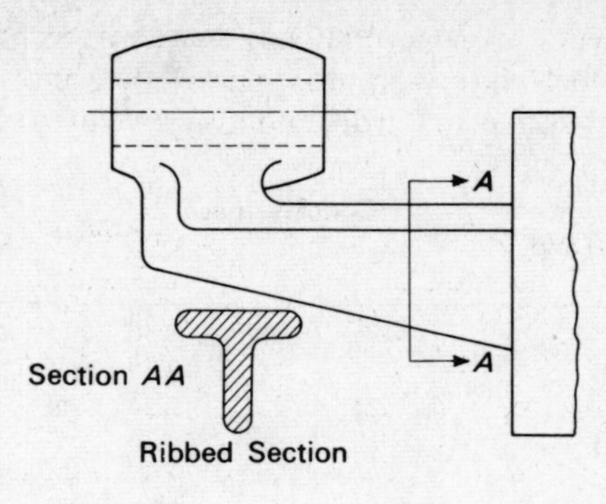

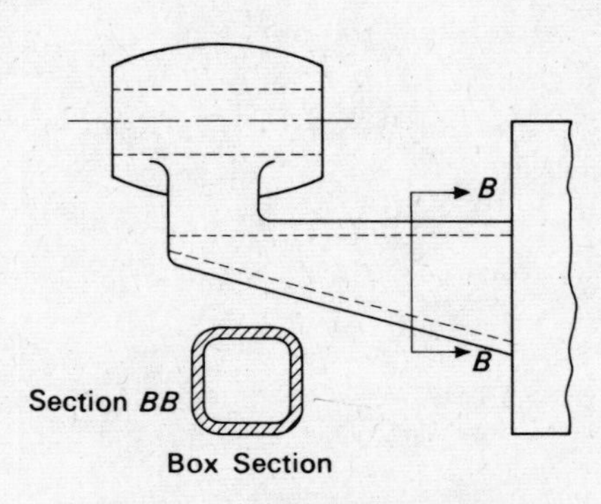

Fig. 7.1 Ribbed and box frames

lathe bed must have large openings for swarf removal, and therefore a ribbed design is used. Plan views of two typical ribbed designs are shown in Fig. 7.3.

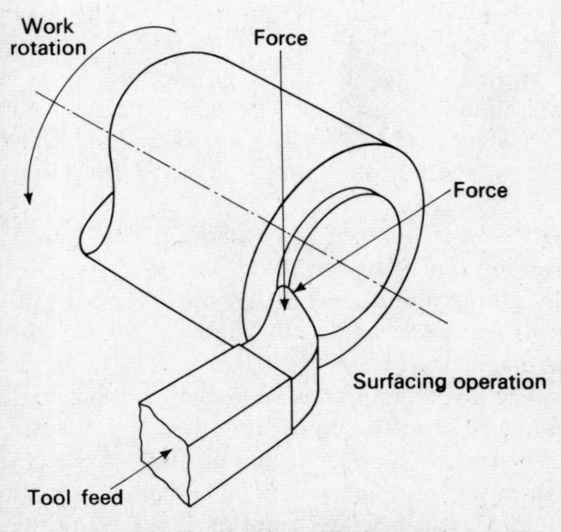

Fig. 7.2 Cutting forces during turning

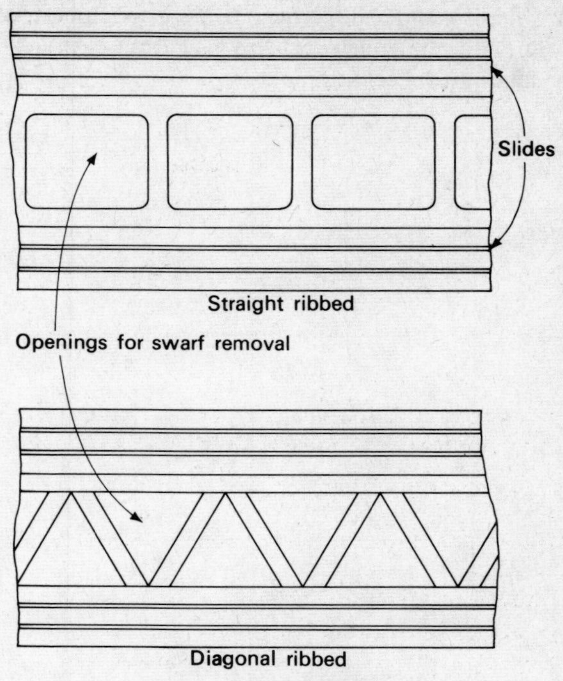

Fig. 7.3 Ribbed lathe beds

The diagonal-ribbed lathe bed has largely replaced the older straight-ribbed design, as it gives a considerably increased resistance to twisting.

All milling machines have a body of box section, as shown in Fig. 7.4.

VIBRATION. During any machining process, the cutting force will vary and tend to set up vibrations in the frame and other parts of the machine. This effect will be magnified if considerable wear exists between the various machine parts. The result of vibrations may well be the production of unsatisfactory work, due to poor surface finish or inaccurate dimensions. Other causes of vibration include faulty gear meshing, joints in driving belts, electric motors, and lack of balance in rotating parts.

A well-designed machine tool frame will absorb vibrations provided it is suitably mounted on solid foundations. Often foundation bolts such as those shown in Fig. 7.5 are used to secure the machine frame to the workshop floor.

CONSTRUCTION. A machine-tool frame may be produced by two basic methods. It may be cast or it may be manufactured in separate sections which are finally joined together. Thus, a machine-tool frame may be constructed in cast iron, cast steel, or fabricated steel sections.

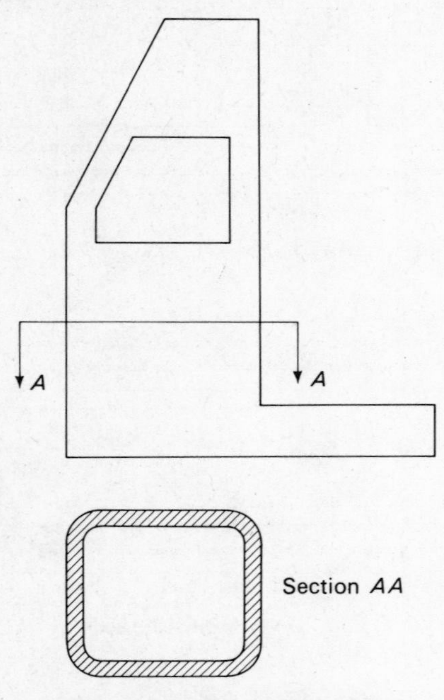

Fig. 7.4 Milling machine body

(a) *Cast iron.* This material is the most widely used for machine-tool frames. Iron castings are relatively inexpensive and can be produced with intricate shapes and sections. Cast iron is strong in compression and, suitably employed, can carry heavy loads with little risk of distortion. It has good machinability and vibration damping properties, and requires minimum lubrication when used as a bearing surface. (See section 7.2.)

(b) *Cast Steel.* Steel castings are much stronger than iron castings, particularly when subjected to tensile (stretching) forces. In addition, they have twice the stiffness and far more resistance to sudden or *shock* loads.

However, the production of steel castings, which requires two annealings, is much more expensive than that of similar iron castings. For this reason, steel castings are rarely used for standard machine tool frames.

(c) *Fabricated steel sections.* The construction of machine-tool frames by welding together separate steel sections is a relatively modern technique. Mild steel plate is approximately two and a half times as stiff as cast iron, and may give up to 50 per cent saving in weight over an equivalent iron casting.

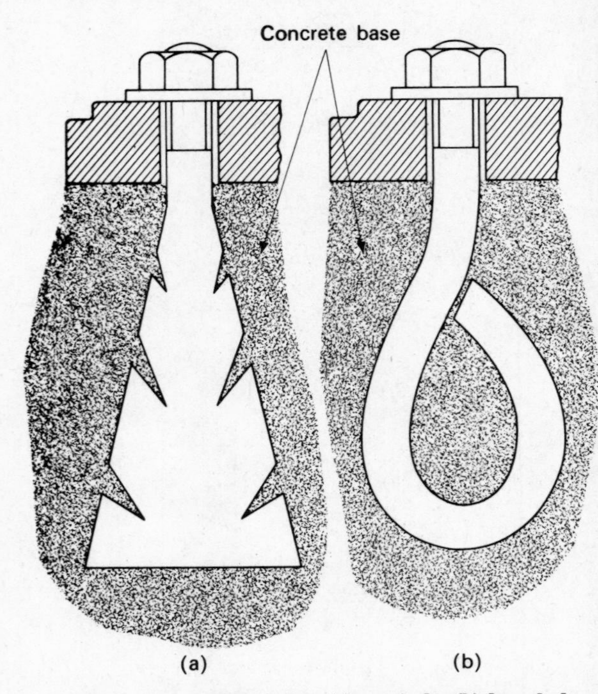

Fig. 7.5 Foundation bolts: (a) rag bolt; (b) loop bolt

High strength and rigidity of steel sections have made it possible to produce machine tool frames which do not rely on solid foundations for the prevention of distortion and vibration effects. In particular, manufacturers of grinding machines have found it possible to produce machines with fabricated steel beds which operate satisfactorily without special foundations or any bolting down.

MOTION AND POWER TRANSMISSION. The modern machine tool is driven by its own electric motor. The power supplied by the motor is carried to the machine-tool spindle through a transmission system.

The power-transmission system of a horizontal milling machine is shown simplified in Fig. 7.6.

The motor supplies rotational power, which is transmitted through belts and pulleys, gears, and a clutch, to provide rotation of the machine spindle. The basic parts of this transmission system are common to all machine tools and are worthy of further consideration.

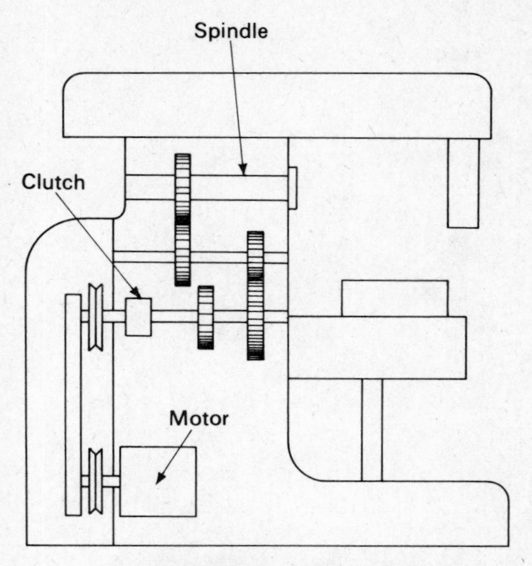

Fig. 7.6 Power transmission

THE ELECTRIC MOTOR. A mains-supplied electric motor provides the driving power required by a machine tool. The design of the motor is such as to cause rotation of a central spindle or *armature.* This spindle may carry an output pulley forming the first stage of the transmission system.

It is suggested that the student should investigate a typical milling machine, and observe how the motor is mounted to allow for adjustment of its position when fitting pulley driving belts. He should also note whether or not the machine-tool motor is automatically isolated from the mains power supply, when the transmission cover is removed. In the case of a satisfactory machine tool, this safety precaution will be provided by a micro-switch, or isolator, working on the principle shown in Fig. 7.7.

PULLEY AND DRIVING BELTS. A common method of transferring motion from one position to another, is the use of a driving belt connecting two pulleys.

A pulley is a circular metal disc which is keyed to a spindle. Pulleys used in machine-tool transmission systems are generally manufactured from cast iron or cast steel, the latter being best suited to high-speed applications.

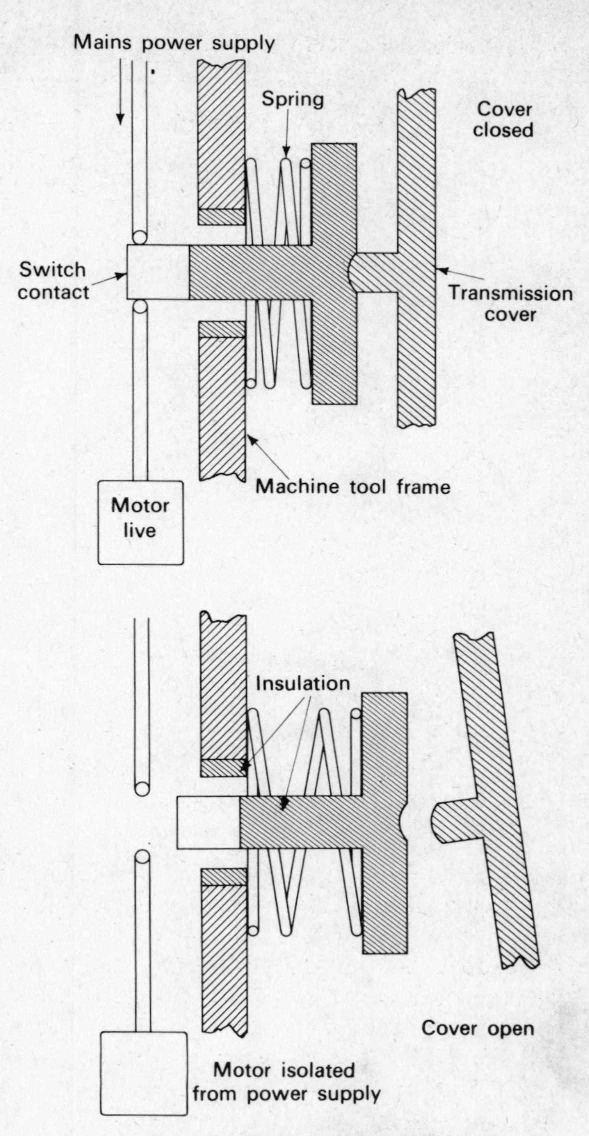

Fig. 7.7 Motor isolation switch

A driving belt may be manufactured from leather, rubber or special fabrics. Three types of belt drives are shown in Fig. 7.8.

(a) *Flat belt drives.* Leather belting is ideal, but more recently interwoven cotton belting is employed to provide a range of speeds using cone pulleys as shown in Fig. 7.9.

The main disadvantage of flat belt drives is that it is difficult to prevent a certain amount of slip from occurring and hence the exact speed of the

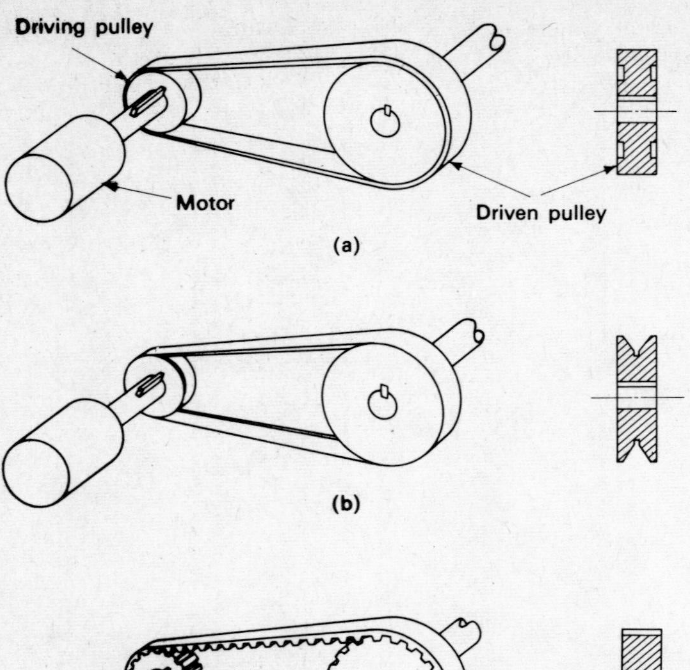

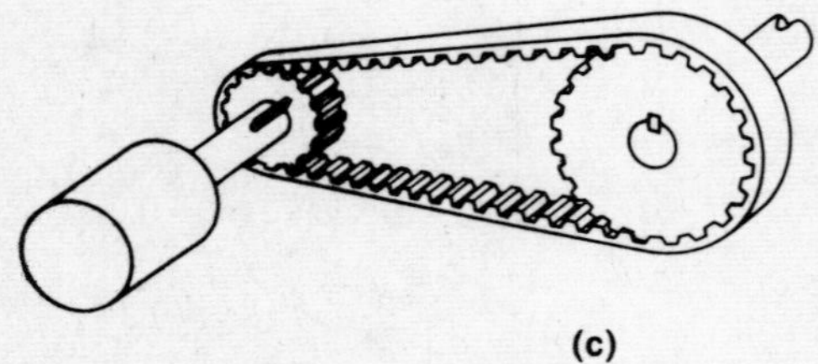

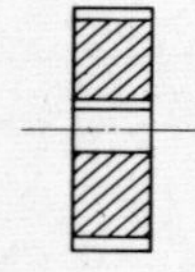

Fig. 7.8 Belt drives: (a) flat; (b) vee; (c) toothed

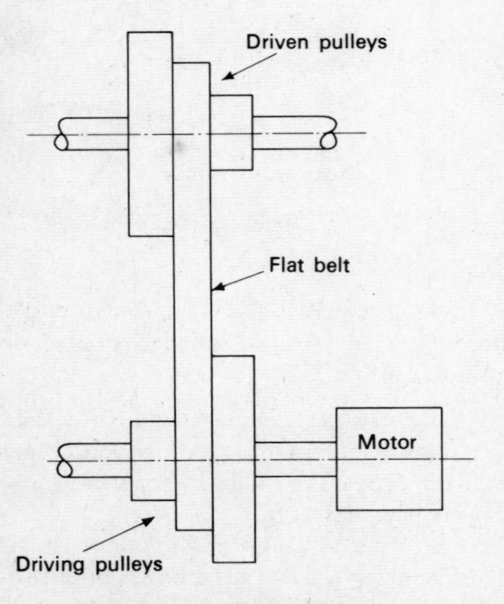

Fig. 7.9 Cone pulleys

driven pulley cannot be guaranteed. For this reason, flat belt drives are not satisfactory for transmission in most feed mechanisms.

(b) *Vee belt drives.* Vee belts are manufactured from rubber or rubber and cotton fabrics. This type of drive is widely used for transmitting motion from the electric motor to the spindle of a machine tool. The shape of the belt section and pulley is shown in Fig. 7.10.

This design provides a wedge action which tends to reduce the problem of belt slip, especially when the driving and driven pulleys are mounted close to each other.

(c) *Tooth belt drives.* Toothed belts are usually manufactured from artificial rubber and covered with a nylon facing. Since the drive is transmitted by belt teeth, the relative speeds of the driving and driven pulleys or gears can be accurately maintained. This type of drive system is very quiet when

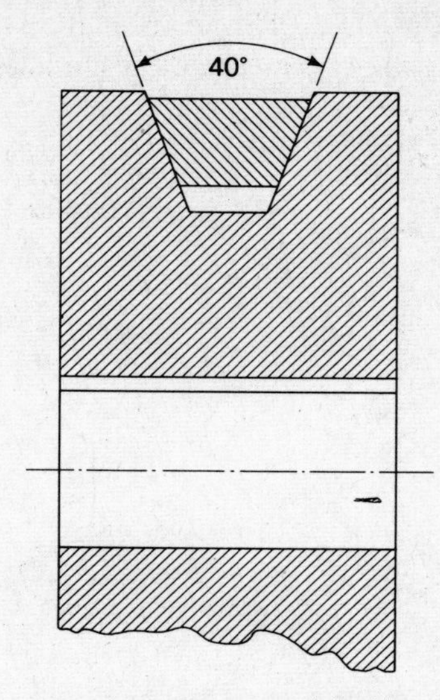

Fig. 7.10 Vee belt and pulley

operating at low speeds, and no lubrication is required.

With all types of belt-drive systems, *endless* belts are preferable to those which are joined, as the latter tend to cause vibrations.

GEARS. The transmission of motion by gears is a positive drive system, in that it provides a definite speed relationship between driving and driven gears. (See Fig. 8.7.) For this reason, gear drives are widely used for feed mechanisms in machine tools.

The range of spindle speeds for centre lathes and milling machines is usually provided by gear systems in modern machine tools. The main reasons for this are:

(a) Accurate values of speed
(b) Ease of speed selection
(c) A wide range of speeds is possible.

A gear drive device worthy of special mention is the *Norton gearbox*. This provides a range of speed relationships between the work spindle and lead-screw of a modern centre lathe.

The Norton gearbox. This component feature of a centre lathe enables the rapid selection of a suitable *gear-train* for screw-cutting and various feeds. The principle of the Norton gearbox is illustrated in Fig. 7.11.

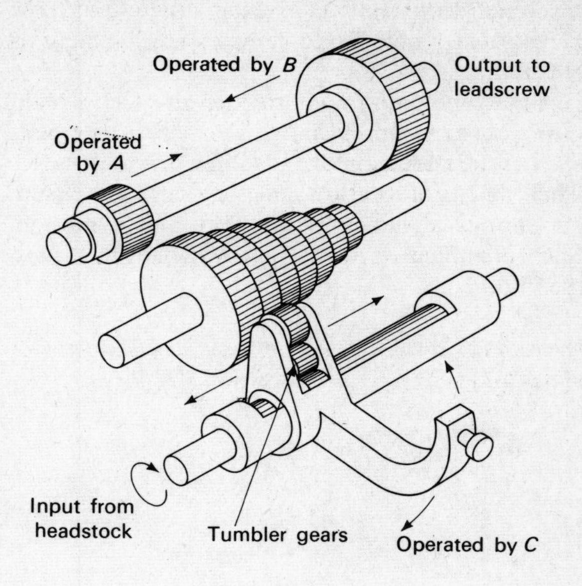

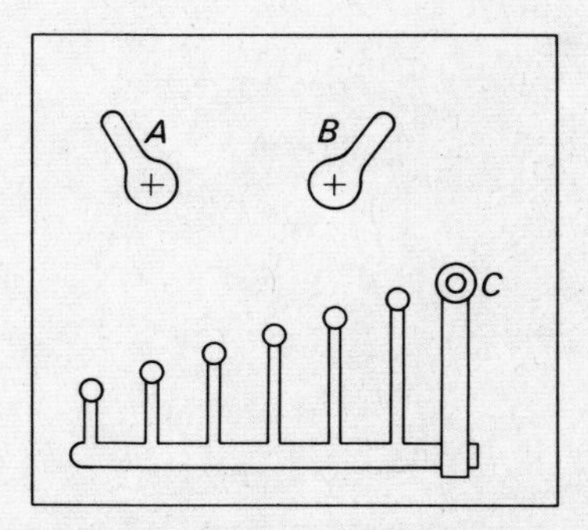

Fig. 7.11 The Norton gearbox

The sliding change gears may be engaged by twisting the selector arms *A* and *B*. The tumbler gears may be engaged at any required gear ratio by sideways movement and lifting of the handle *C*.

The smaller sliding gear may be engaged, with the tumblers engaging any one of seven cone gears. Similarly seven gear ratios may be obtained using the larger sliding gear.

Therefore, fourteen different screwthreads may be cut without the need for manual removal and changing of gearwheels.

7.2 Alignment in machine tools

Having decided that a machine tool frame must be rigid to provide accurate location of other machine parts relative to each other, it is now necessary to investigate how their alignment is ensured.

A machine tool frame provides, in effect, a main datum from which all moving machine members, such as spindles, saddles and tables, may be located. Their means of location must be such as to ensure continuous alignment when they are in motion. Such alignment is usually provided by bearings. (See Vol. 2.)

SLIDES AND SLIDEWAYS. The basic design forms of slides and slideways are shown in Fig. 7.12.

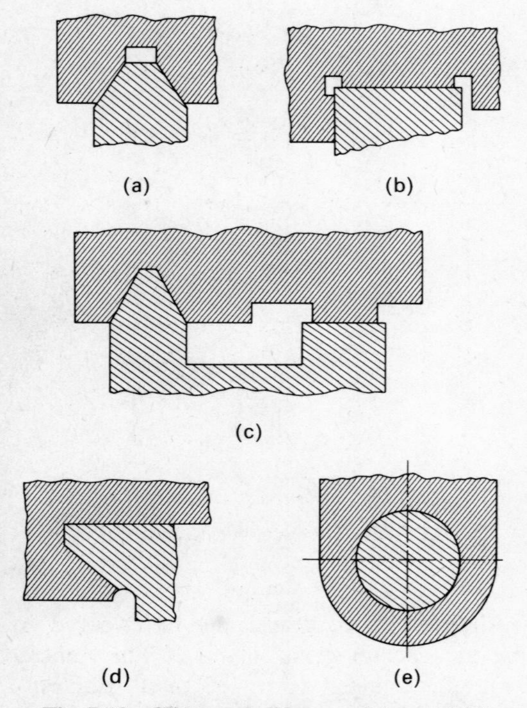

Fig. 7.12 Slides and slideways: (a) vee; (b) flat; (c) vee and flat combination; (d) dovetail; (e) cylindrical

(a) *Vee slides.* This design provides accurate location in both horizontal and vertical directions. The upward vee form also prevents swarf from accumulating on the slideways.

(b) *Flat slides.* These are employed mainly for heavy-duty work and large machine tables.

(c) *Vee and flat slides.* This design is widely used on lathe beds, for saddle and tailstock alignment, as shown in Fig. 7.13. This combination is highly effective as contact between three surfaces is easily maintained.

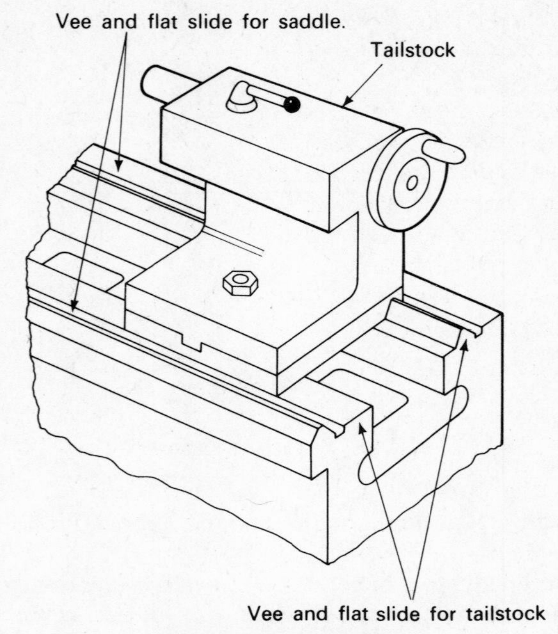

Fig. 7.13 Lathe-bed slides

(d) *Dovetail slides.* These are widely used to guide machine-tool members producing feed movements, such as lathe cross slides, milling-machine tables, and shaper head slides. Figure 7.14 illustrates the application of dovetail slides and slideways

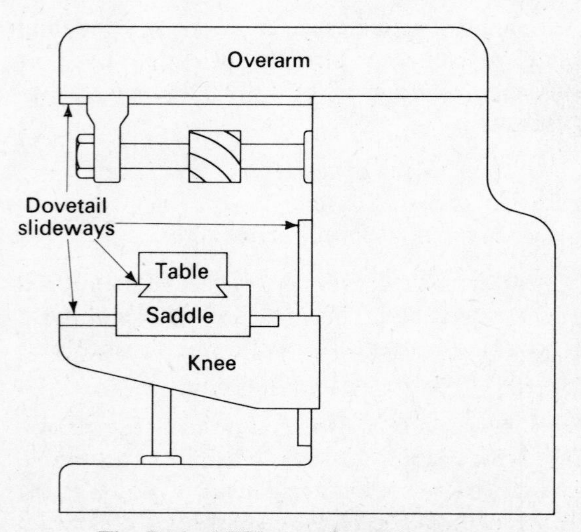

Fig. 7.14 Milling machine slideways

to the overarm, table, saddle and knee of a horizontal milling machine.

(e) *Cylindrical slides.* These provide very accurate alignment and are often employed in the design of boring and drilling machines.

WEAR ADJUSTMENT. To prevent vibration and maintain accurate alignment, slides and slideways are usually provided with some means of wear adjustment. It is suggested that the student should investigate centre lathes and milling machines to observe means of adjustment in slideways. A typical method of adjusting a dovetail slideway is shown in Fig. 7.15.

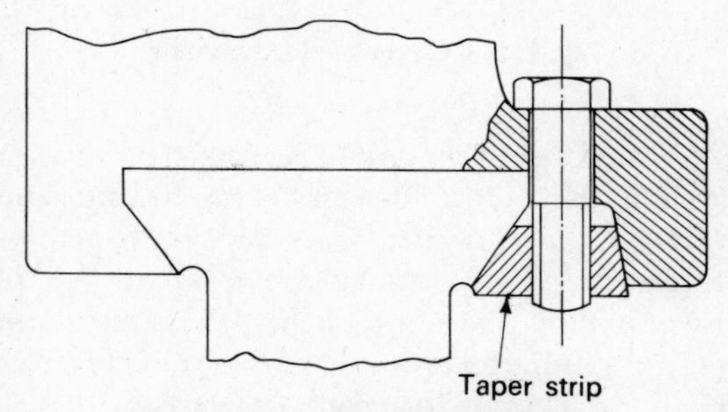

Fig. 7.15 Wear adjustment

CHECKING MACHINE TOOL ALIGNMENTS. Machine tool alignments should be checked at frequent intervals, to ensure the production of accurate work. A dial test indicator may usefully be employed for such an investigation. It is suggested that as a workshop activity the student should list the principal alignments of a particular machine tool and carry out a series of detail checks. All results should be recorded, together with recommendations for correcting any possible sources of error.

7.3 Relative movement between cutting tool and workpiece

When workpieces are produced by machine-tool processes, such as turning and milling, their manufacture may be considered simply as a series of *forming* and *generating* operations.

FORMING. Any production process where the shape of the tool produces an identical reverse form in the workpiece is termed a *forming process.* The tool employed is termed a *form tool.*

Figure 7.16 shows an undercut being turned using a form tool. This is a forming process, as no other shape of tool would produce the same result.

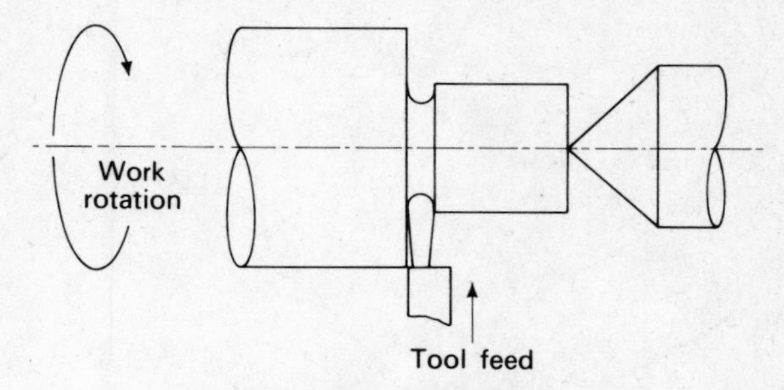

Fig. 7.16 Forming

GENERATING. A generating process is one where the resulting shape of the workpiece is not directly related to the shape of the cutting tool. The shape produced is, in fact, entirely due to the relative motion between the cutting tool and workpiece.

Figure 7.17 shows a cylindrical feature being turned. The shape produced is due to the rotation of the work and to the traverse of the tool parallel to the work axis.

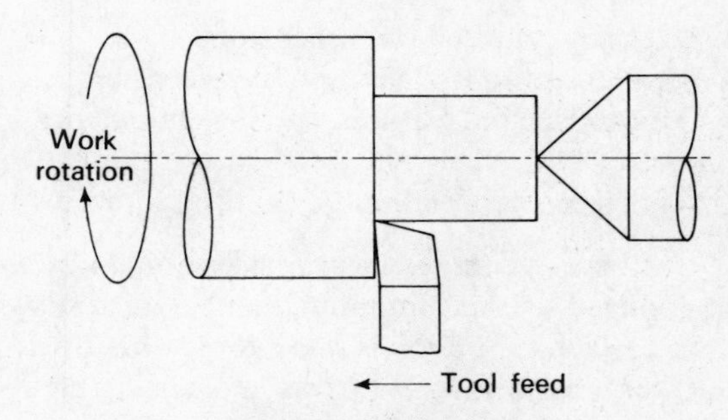

Fig. 7.17 Generating

Figure 7.18 illustrates the production of a flat surface by generating. The machine tool indicated is a shaper, but the principle involved is equally applicable to a milling, grinding or planing machine tool.

WORKPIECE PRODUCTION — PRELIMINARY CONSIDERATIONS. Before the production of any workpiece is started, a sequence of required operations

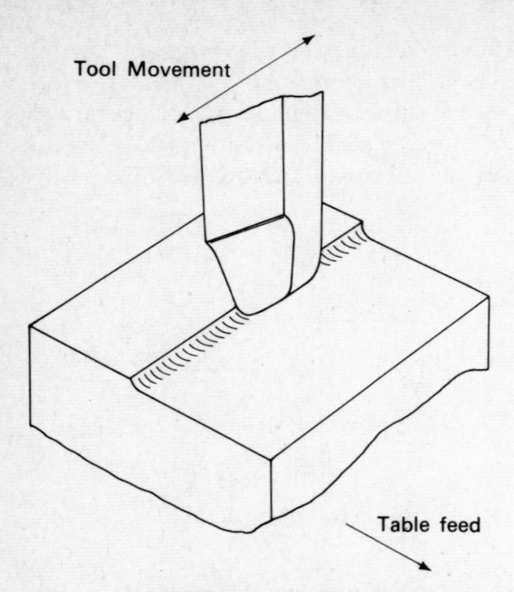

Fig. 7.18 Generating a flat surface

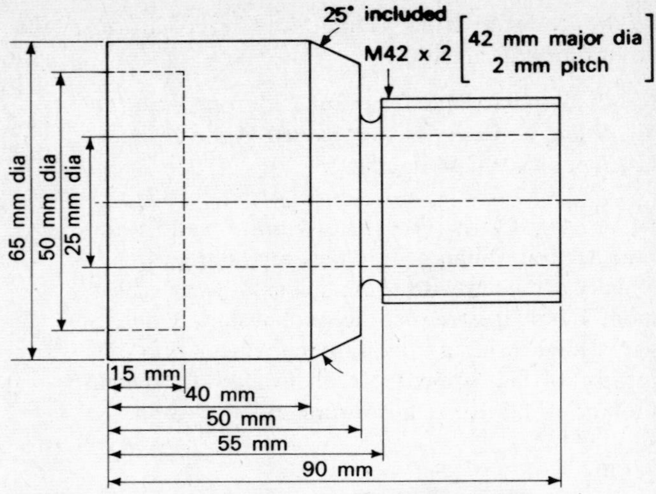

Fig. 7.19 Component for turning

should be planned. This sequence of operations should include suitable methods of work holding and tool setting, to provide the required degree of accuracy. This will entail studying a drawing of the workpiece to allow for such considerations as the following:

(a) Accuracy required on dimensions
(b) Parallelism and squareness between features
(c) Concentricity of features, such as diameters and bores
(d) Workpiece material and quantity required.

To achieve accuracy, a workpiece should always be produced at the minimum number of settings possible. Each time a job is re-set for further operations, the possibility of introducing setting errors arises.

PRODUCTION OF WORKPIECES. To conclude this chapter, the production of three typical workpieces will be considered. Their production will involve the use of:

(a) The centre lathe
(b) The milling machine
(c) The shaper.

(a) *Centre-lathe turning.* The component shown in Fig. 7.19 is to be manufactured on a centre lathe, only one off being required. The material supplied is 70 mm diameter mild steel bar by 100 mm long.

Examination of the workpiece drawing will indicate that two settings are required for the production of this component. The first setting should result in the production of a reliable datum surface for accurate location at the second setting.

During both settings the work must be located on a common centre-line to ensure that all features are concentric. Due to its lack of accuracy, a three-jaw chuck could not be employed for the second setting as concentricity in the workpiece would be lacking. One suitable method of re-setting the workpiece would be by gripping it in a four-jaw chuck, although the use of a mandrel could well be considered. The sequence of operations is as follows:

First setting
Op. 1 Grip the bar in a three-jaw chuck and face one end
Op. 2 Turn the 65 mm diameter for a length of approximately 60 mm
Op. 3 Centre drill, pilot drill and drill the hole 20 mm diameter
Op. 4 Bore the hole to 25 mm diameter
Op. 5 Bore the recess to 50 mm diameter

Second setting
Op. 6 Reverse the workpiece, grip in a four-jaw chuck and set accurately as shown in Fig. 7.20

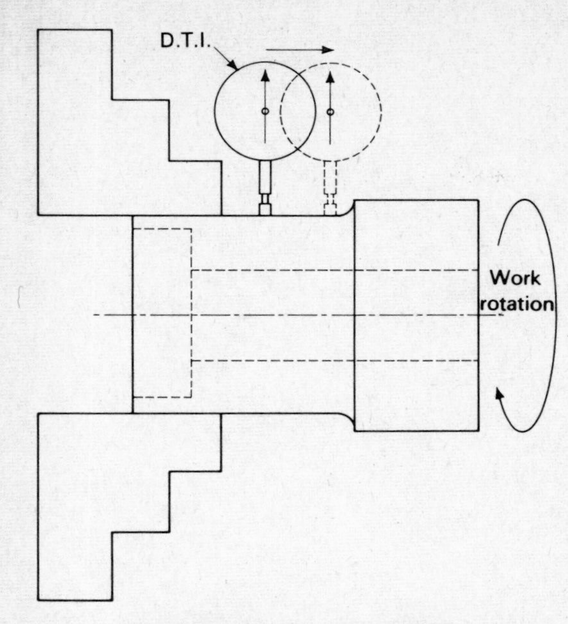

Fig. 7.20 Four-jaw chuck setting

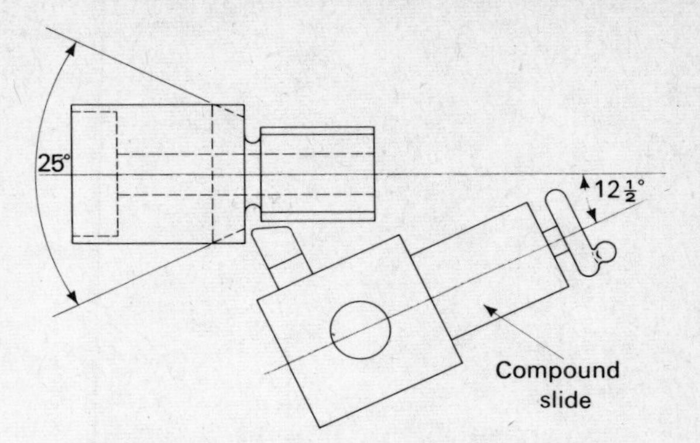

Fig. 7.21 Compound slide set for taper turning

Op. 7 Face workpiece to 90 mm length

Op. 8 Turn the 42 mm diameter for a length of 40 mm. A revolving centre, mounted in the tailstock, should be used to support the workpiece during this and subsequent operations

Op. 9 Turn the undercut using a radius-form tool

Op. 10 Cut the screwthread. This will entail mounting a suitably ground tool square to the workpiece and selecting the required gear ratio in the Norton gearbox

Op. 11 Turn the taper by setting the compound slide at half the included angle required as shown in Fig. 7.21

Op. 12 Remove all burrs and sharp edges.

It is not intended to discuss in detail the processes of screwcutting and taper turning at this stage. It is, however, suggested that the student should investigate alternative methods of carrying out these processes as a supervised workshop activity.

(b) *Milling*. The component shown in Fig. 7.22 is to be completed on a vertical milling machine. The material supplied is mild-steel bar with all turning operations completed.

Examination of the workpiece drawing will suggest that a vertical milling machine and dividing head could well be employed for the completion of this component at one setting.

Before outlining the milling processes required, brief consideration will be given to the operation of a dividing head. The mechanism is shown in simplified form in Fig. 7.23.

One complete turn of the handle will rotate the wormwheel one fortieth of a revolution. As the workpiece chuck is secured to the same spindle as the wormwheel, the workpiece will simultaneously

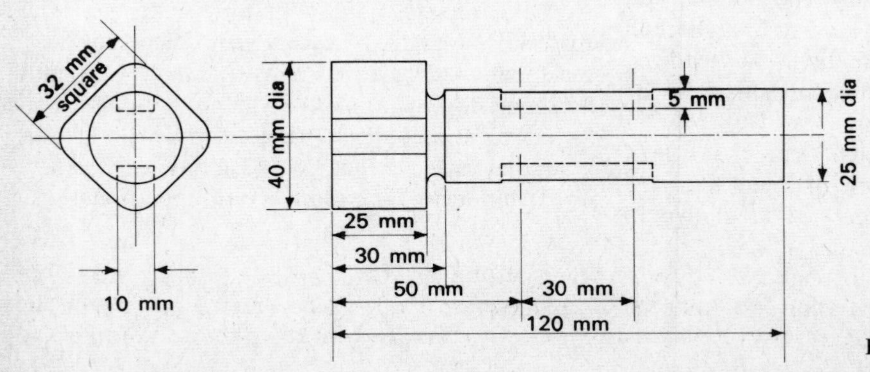

Fig. 7.22 Component for milling

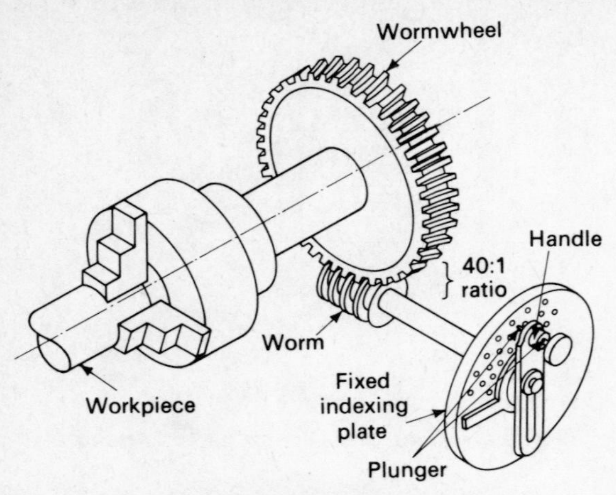

Fig. 7.23 Dividing-head mechanism

be rotated through one fortieth of 360°, i.e., 9°. To index the chuck through less than 9°, the handle must be rotated through less than one complete turn. To permit this, the fixed indexing plate contains several circles of holes and the spring-loaded plunger may locate the handle at any hole during a revolution.

The movement of the handle required for indexing the work through any required angle, may be determined from the following formula:

Movement required = Angle required × 1/9 turns.

It is suggested that this should be proved and applied in practice as part of a workshop activity. More detailed consideration is given to the dividing head in Vol. 2.

The workpiece under discussion could be completed by the following sequence of operations:

Op. 1 Grip the smaller diameter in the dividing-head chuck and support the 'square' end by means of a centre, having first checked the alignment of the dividing head by means of a dial test indicator affixed to the milling machine frame

Op. 2 Using an end mill, machine the first side of the square. The depth of material to be removed can be calculated as shown in Fig. 7.24

Op. 3 Index the workpiece through 90° (10 complete turns of the handle) and mill the second side of the square

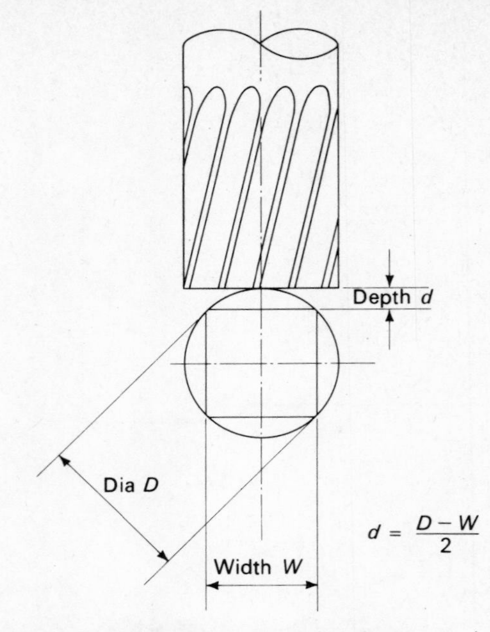

Fig. 7.24 Depth of cut calculation

Op. 4 Repeat Op. 3 twice more to complete the square feature

Op. 5 Index the workpiece through 45° (5 complete turns of the handle)

Op. 6 Mount a 10 mm slot drill in position, using the principle indicated in Fig. 7.25

Op. 7 Mill the first groove 5 mm deep, for a length of 30 mm

Op. 8 Index the workpiece through 180° (20 complete turns of the handle)

Op. 9 Mill the second groove 5 mm deep, for a length of 30 mm

Op. 10 Remove all burrs and sharp edges.

(c) *Shaping.* The component shown in Fig. 7.26 is to be completed on a shaping machine. The material supplied is a cast iron block, already machined square to 65 mm × 90 mm × 100 mm and marked out.

Examination of the workpiece drawing will indicate that two settings are required to complete the component. The sequence of operations is as follows:

First setting

Op. 1 Secure a vice to the shaping machine table, with the vice jaws at 90° to the movement of

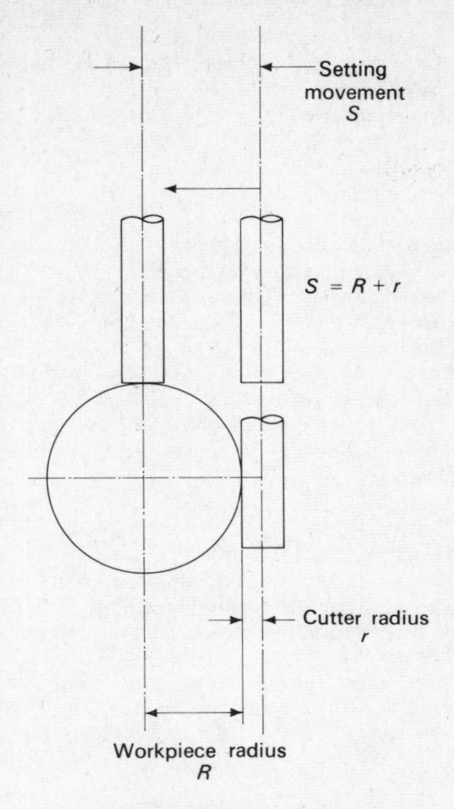

Fig. 7.25 Cutter setting

the ram. Check this alignment, using a dial test indicator held in the machine tool post

Op. 2 Grip the workpiece in the vice, using parallel strips to support the block at a satisfactory height

Op. 3 Using a suitably ground tool, shape out the step to the required depth of 25 mm, using a fine table feed as shown in Fig. 7.27

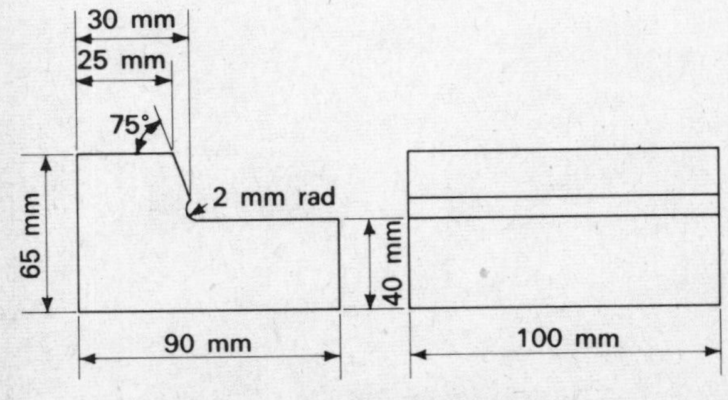

Fig. 7.26 Component for shaping

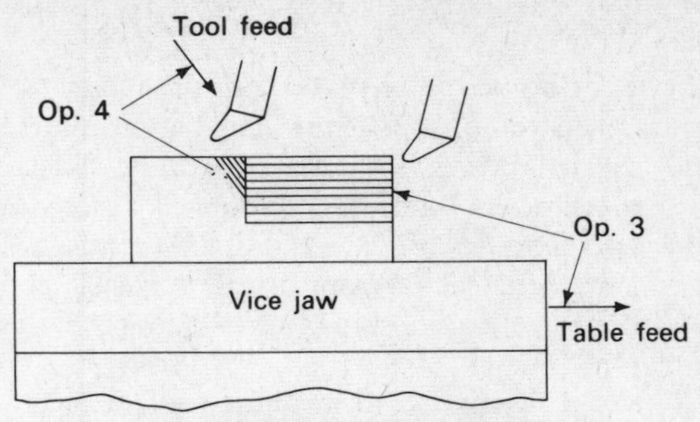

Fig. 7.27 Shaping operations

Second setting

Op. 4 Set the headslide to 15° (90° – 75°) and shape the angular face as shown in Fig. 7.27

Op. 5 Remove all burrs and sharp edges.

Notice that, when shaping angular faces, the headslide is set to the angle required but the swivel plate must be tilted in the opposite direction. This will allow the clapper box to provide tool clearance during the return stroke of the shaper ram. This important factor is illustrated in Fig. 7.28.

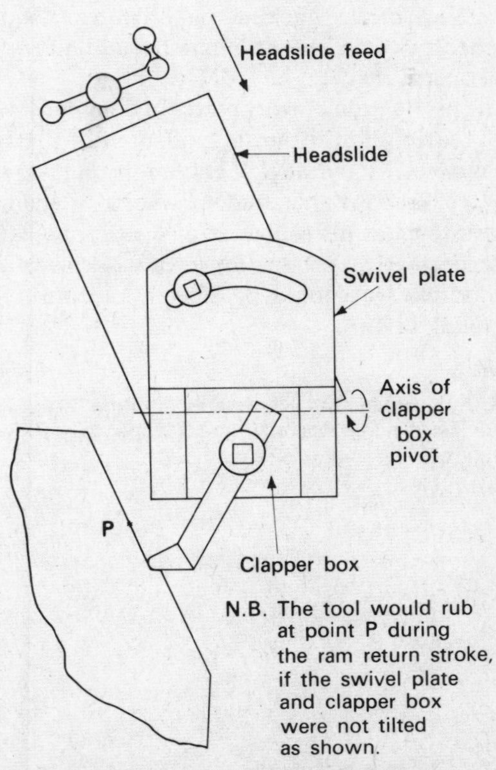

Fig. 7.28 Set-up for shaping angular faces

Summary

The main frame, body or bed of a machine tool is designed to resist twisting and bending forces and also to absorb vibrations. The most satisfactory machine-tool frames have *box* or *diagonal ribbed* sections to provide rigidity. Machine-tool frames are constructed from cast iron, cast steel or fabricated steel sections, the former being the most common.

Power transmission in machine tools is usually provided by pulleys and driving belts, or systems of gears. Driving belts may be of *flat*, *vee* or *toothed* form, manufactured from leather, rubber or special fabrics. Gear transmissions are widely used for feed mechanisms, where accurate motions are required, and to provide a range of speeds for machine-tool spindles.

A machine-tool frame provides a *main datum* from which all moving members, such as spindles, saddles and tables, may be located. Slides and slideways serve to provide such location, and to maintain alignment. The most common types of slides and slideways have vee, flat, dovetail or cylindrical forms. Machine-tool alignments should be checked at frequent intervals and necessary adjustments made.

The production of workpieces of different shapes will involve control by the craftsman of the relative movement and angle between the cutting tool and workpiece. The student should remember always to plan his sequence of operations before commencing the production of a workpiece. The method chosen should be efficient, accurate, but above all, safe.

Questions

1. With the aid of sketches, describe the *basic structure* of a pillar drilling machine, and explain the following features:
   (a) How a main datum is provided
   (b) How alignment between the main features is ensured.
2. Sketch and describe three uses of each of the following in a machine shop:
   (a) Flat belts and pulleys
   (b) Vee belts and pulleys
   (c) Gear trains.
3. Sketch and describe two applications of dovetail slides and slideways on a shaping machine.
4. List five important alignments on a centre lathe, and describe, with the aid of diagrams, how each could be checked.
5. Sketch and describe three different methods of wear adjustment used on machine tools.
6. State one advantage and one disadvantage of using each of the following for the construction of machine-tool frames.
   (a) Cast iron
   (b) Cast steel
   (c) Fabricated steel sections.
7. By means of line diagrams, show three methods of generating a flat surface by machining. In each case show the relative movement between the tool and workpiece.
8. Describe, illustrating a sequence of operations with sketches, how the component shown in Fig. 7.29 could be produced.

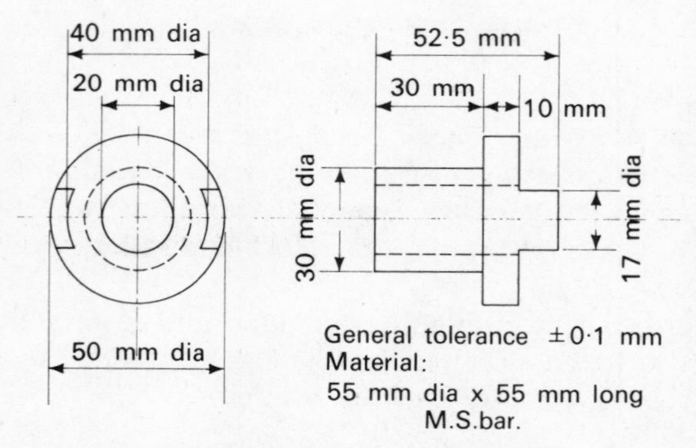

Fig. 7.29 Coupling component

8. Studies associated with machine tools

This chapter examines the following topics associated with machine tools:

8.1 Friction as a driving force
8.2 Gear-drive calculations
8.3 Pulley-drive calculations
8.4 Force-ratio and movement-ratio calculations.

8.1 Friction as a driving force

In section 6.2, friction was considered as a force which tends to resist sliding motion between two surfaces in contact. It was found that friction was not only an advantage, but a necessity in clamping devices.

In the same manner, friction may be used to advantage as a driving force, by preventing two surfaces in contact from slipping.

This principle is employed in the following driving devices, associated with machine tools:

(a) Driving belts and pulleys
(b) Clutches
(c) Tool shanks.

(a) FRICTION BETWEEN DRIVING BELTS AND PULLEYS. In section 7.1, three types of driving belts and pulleys were discussed. Two of these—namely, flat-belt and vee-belt drives—depend on friction between the belt and pulley surfaces to provide a driving action. *Slip* will occur if the friction force between the surfaces in contact is lost.

In order to maintain friction, suitable materials are employed for the pulleys and belt. A leather-soled shoe will slip on an ice-covered road due to lack of friction. This effect would not occur on a dry road surface with a rubber-soled shoe. Hence, the material nature of the surfaces in contact will decide the size of the friction force between them. Leather- and rubber-based fabrics create large friction forces when moving in contact with steel or cast iron surfaces. Combinations of these materials are, therefore, used for driving belts and pulleys.

In Fig. 8.1 a belt is shown driving a pulley by three methods. With a little imaginative thought, the following conclusions should be reached.

With method (A), there is a considerable possibility that the belt will slip and the pulley lose speed. This effect is less likely with method (B), and most unlikely with method (C).

Hence it is true to say that the driving force of friction is at its greatest value when the arc of contact between the belt and pulley is at a maximum.

Figure 8.2 illustrates a horizontal belt drive system rotating such that:

(a) the tight side of the belt is at the bottom
(b) the tight side of the belt is at the top.

Notice at (b), when the slack of the belt is at the bottom it tends to drop away from the pulleys. This decreases the arcs of contact and thereby reduces the frictional driving force. For this reason, horizontal belt drives should always be arranged as shown at (a).

With vertical belt-drive systems, there is a tendency for the belt to drop away from the lower pulley. This can be prevented by the addition of a *jockey* pulley or tensioner as shown in Fig. 8.3.

(b) FRICTION CLUTCHES. A clutch is a means of connecting two shafts such that the drive between them may be disconnected at will.

In its simplest form, a *friction* clutch consists of two circular metal plates mounted on adjoining

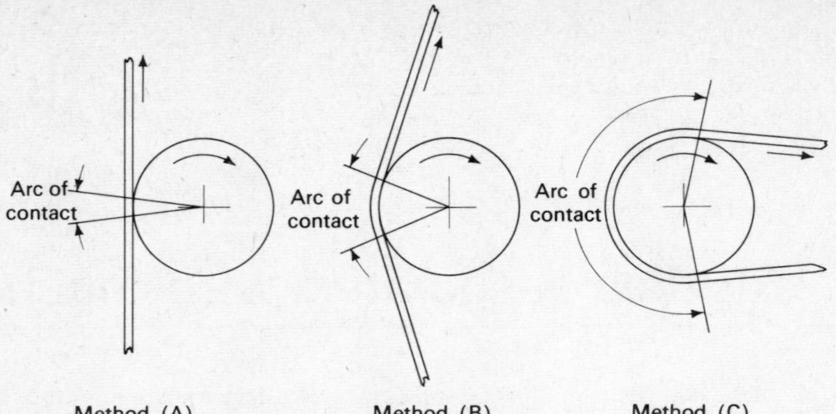

Fig. 8.1 Belt driving Method (A) Method (B) Method (C)

shafts. Each plate is keyed to its shaft, one tightly and the other loosely, such that it may slide along the shaft. A disc of high friction material is riveted or bonded to the inner face of the fixed or driving plate, as shown in Fig. 8.4.

When the sliding plate is made to contact the driving plate, the friction force between them will cause the former to rotate. In this manner, rotational power is transmitted between the shafts.

In order to assist the friction force, a spring is often employed to increase the pressure between

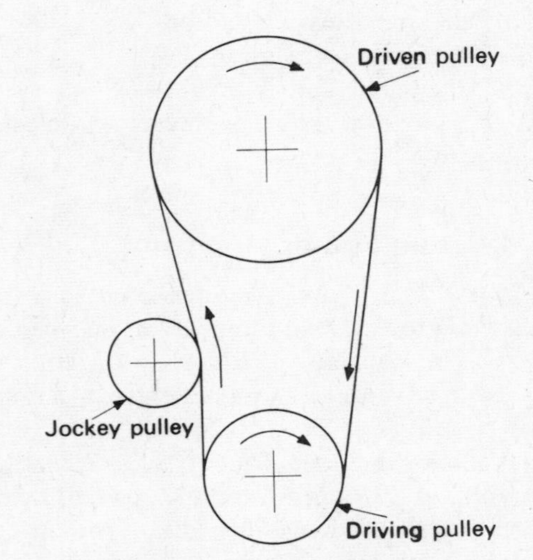

Fig. 8.3 Use of jockey pulley

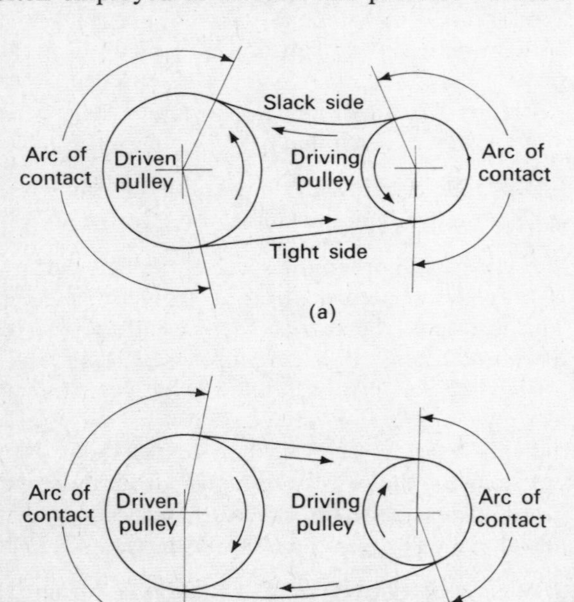

Fig. 8.2 Horizontal belt drives

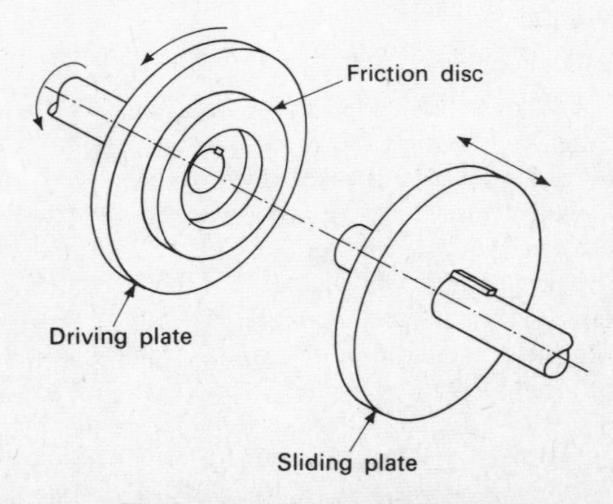

Fig. 8.4 Friction clutch principle

the two plates. A simple clutch assembly is shown sectioned in Fig. 8.5.

This principle is widely employed in machine-tool transmission systems but the designs of clutches used are many and varied.

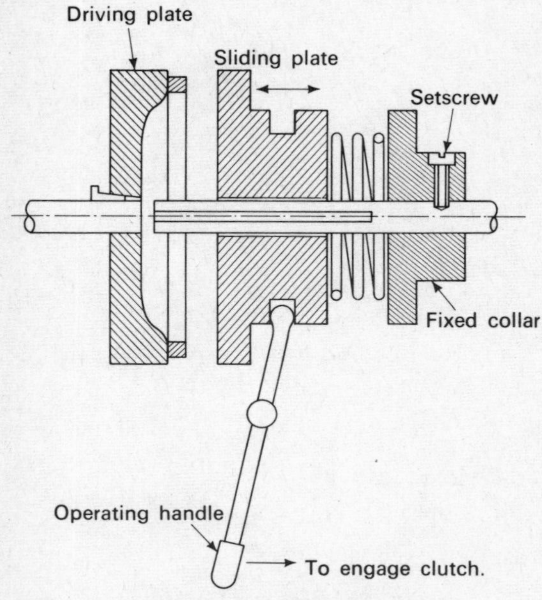

Fig. 8.5 Simple clutch assembly

(c) FRICTION-DRIVEN TOOL SHANKS. Many cutting tools, such as drills and machine reamers, have tapered shanks designed to fit into a machine-tool spindle as shown in Fig. 8.6.

The angle of the taper employed ensures that the tool may be driven due to friction between the shank and spindle surfaces.

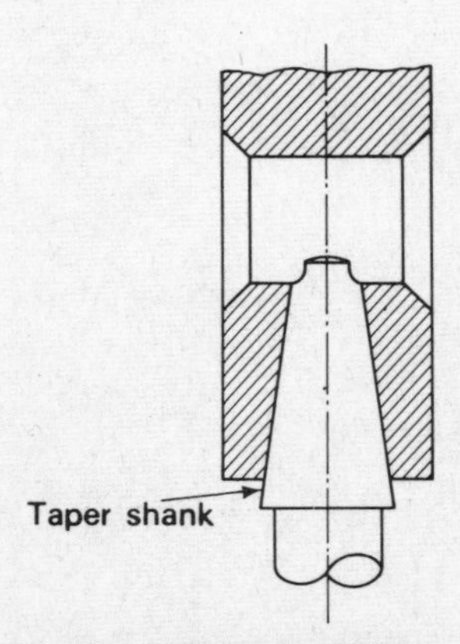

Fig. 8.6 Taper shank drive

8.2 Gear-drive calculations

In section 7.1 gear drives were mentioned as a means of power transmissions. Figure 8.7 illustrates a gear-drive system between the spindle and leadscrew of a centre lathe.

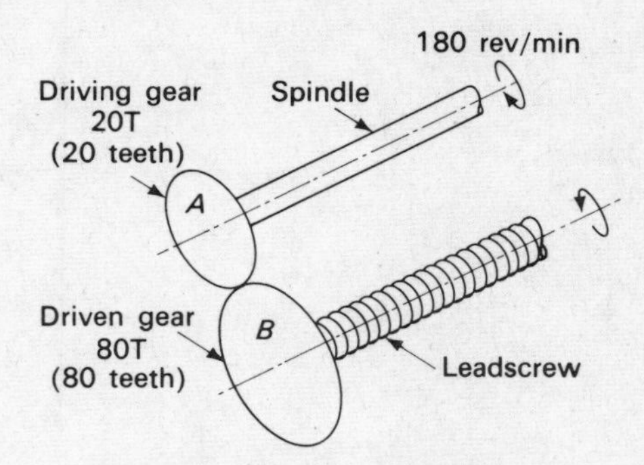

Fig. 8.7 Gear drive system

The rotational speed of the leadscrew can be calculated as follows:

$$\frac{T_A}{T_B} = \frac{\text{speed of driven gear}}{\text{speed of driving gear}}$$

$$\frac{(\text{rev/min})}{(\text{rev/min})} \quad \text{or} \quad \frac{(\text{rev/s})}{(\text{rev/s})}$$

where

T_A = number of teeth in gear A
T_B = number of teeth in gear B

$$\therefore \quad \frac{20}{80} = \frac{\text{speed of driven gear}}{180}(\text{rev/min})$$

$$\therefore \quad \frac{20 \times 180}{80} = \text{speed of driven gear}$$

$$= 45 \text{ rev/min}$$

$\therefore$ rotational speed of leadscrew = 45 rev/min.

To reverse the direction of rotation of the driven gear, a third gear, called an *idler*, is meshed between the driving and driven gears. The number of teeth in an idler gear does not affect the speed calculation. Gear drives of this type are called *simple* gear trains, each gear being fixed to its own shaft.

The gear drive shown in Fig. 8.8 is called a *compound* gear train.

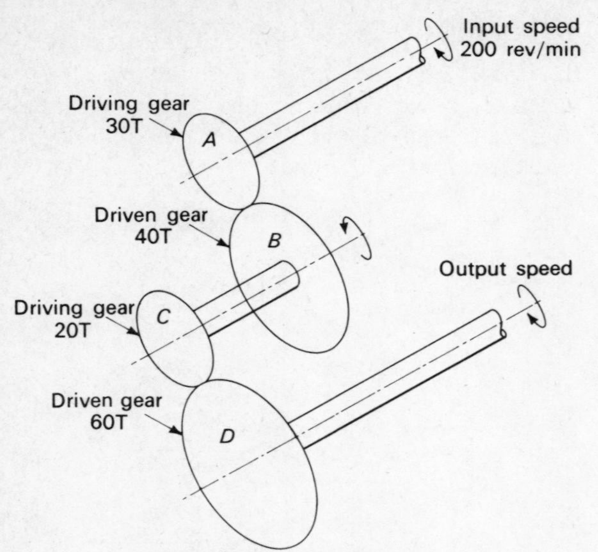

Fig. 8.8 Compound gear train

The output speed of this gear train can be calculated as follows:

$$\frac{T_A}{T_B} \times \frac{T_C}{T_D} = \frac{\text{output speed}}{\text{input speed}}$$

$$\frac{(\text{rev/min})}{(\text{rev/min})} \quad \text{or} \quad \frac{(\text{rev/s})}{(\text{rev/s})}$$

where

T_A = number of teeth in gear A
T_B = number of teeth in gear B
T_C = number of teeth in gear C
T_D = number of teeth in gear D

$$\therefore \quad \frac{30}{40} \times \frac{20}{60} = \frac{\text{output speed}}{200} \qquad (\text{rev/min})$$

$$\therefore \quad \frac{30}{40} \times \frac{20}{60} \times 200 = \underline{\text{output speed} = 50 \text{ rev/min.}}$$

The application of the calculations above is shown in the following examples:

Example 1 If the leadscrew of a lathe has a lead of 6 mm and the meshing gears in the quick-change (Norton type) gearbox for cutting a lead of 2 mm are 14 T driver and 33 T driven, determine the ratio of the 'pick-off' gearing required between the lathe spindle and quick-change gearbox.

This arrangement is shown in Fig. 8.9.

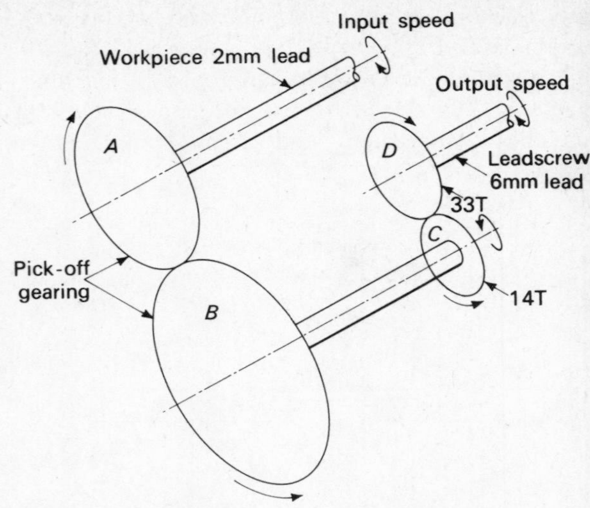

Fig. 8.9 Gear train for screw cutting

Now,

$$\frac{\text{output speed}}{\text{input speed}} = \frac{T_A}{T_B} \times \frac{T_C}{T_D}$$

where

T_A = number of teeth in gear A
T_B = number of teeth in gear B
T_C = number of teeth in gear C
T_D = number of teeth in gear D.

But

$$\frac{\text{output speed}}{\text{input speed}} = \frac{\text{lead of work}}{\text{lead of leadscrew}} = \frac{2}{6}.$$

This means the lathe spindle (input) must rotate three times as fast as the leadscrew in order to produce a thread three times as fine as the leadscrew thread.

Therefore

$$\frac{\text{output speed}}{\text{input speed}} = \frac{2}{6} = \frac{T_A}{T_B} \times \frac{T_C}{T_D}$$

$$\therefore \quad \frac{2}{6} = \frac{T_A}{T_B} \times \frac{14}{33}$$

$$\therefore \quad \frac{2}{6} \times \frac{33}{14} = \frac{T_A}{T_B} = \frac{11}{14}$$

$\therefore$ $\underline{\text{ratio of the pick-off gearing} = 11:14.}$

Example 2 The all-geared headstock of a certain centre lathe is shown in Fig. 8.10, together with information concerning its gears.

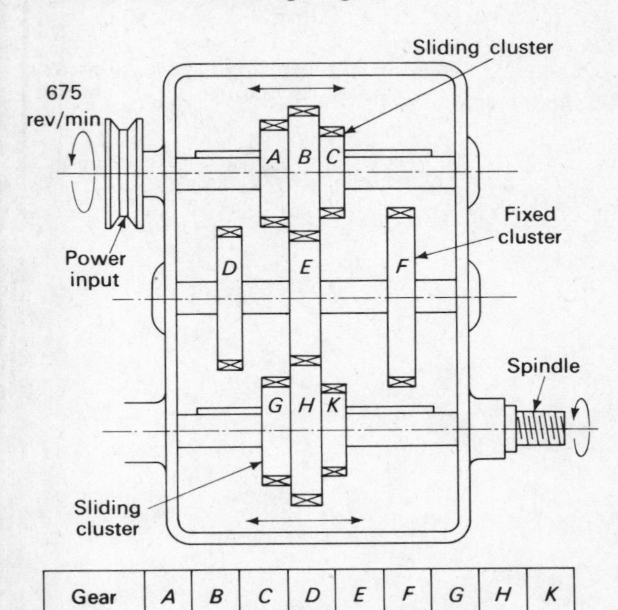

Gear	A	B	C	D	E	F	G	H	K
Number of teeth	55	70	40	85	70	100	60	75	45

Fig. 8.10 All-geared headstock

(a) State the number of speeds available
(b) Calculate the spindle speed for the gear arrangement shown
(c) Calculate the lowest spindle speed available
(d) Calculate the highest spindle speed available.

(a) The arrangement allows for the use of any one of three input driving gears A, B, or C. At the same time it permits the use of any one of three output driven gears G, H, or K.

Therefore, the number of possible combinations will be:

$$3 \text{ inputs} \times 3 \text{ outputs} = \underline{9 \text{ speeds.}}$$

The nine gear trains producing these speeds are as follows:

(1) $\dfrac{A}{G}$ using D as an idler (simple gear train)

(2) $\dfrac{A}{D} \times \dfrac{E}{H}$ (compound gear train)

(3) $\dfrac{A}{D} \times \dfrac{F}{K}$ (compound gear train)

(4) $\dfrac{B}{H}$ using E as an idler (simple gear train)

(5) $\dfrac{B}{E} \times \dfrac{D}{G}$ (compound gear train)

(6) $\dfrac{B}{E} \times \dfrac{F}{K}$ (compound gear train)

(7) $\dfrac{C}{K}$ using F as an idler (simple gear train)

(8) $\dfrac{C}{F} \times \dfrac{D}{G}$ (compound gear train)

(9) $\dfrac{C}{F} \times \dfrac{E}{H}$ (compound gear train)

(b) The arrangement shown in Fig. 8.10 is a simple gear train, and the idler gear E does not affect the calculation of the spindle speed.

Therefore

$$\frac{T_B}{T_H} = \frac{\text{speed of driven gear}}{\text{speed of driving gear}} \quad \frac{(\text{rev/min})}{(\text{rev/min})}$$

$$\therefore \quad \frac{70}{75} = \frac{\text{speed of driven gear}}{675}$$

$$\therefore \quad \frac{70}{75} \times 675 = \text{speed of driven gear}$$

$$= \underline{630 \text{ rev/min} = \text{spindle speed.}}$$

(c) When any gearbox permits the selection of both simple and compound gear trains, the highest and lowest output speeds will be obtained using compound gear trains.

The lowest speed available will employ the smallest driving gears in engagement with the largest driven gears, as shown in Fig. 8.11.

Therefore

$$\frac{T_C}{T_F} \times \frac{T_E}{T_H} = \frac{\text{output speed}}{\text{input speed}} \quad \frac{(\text{rev/min})}{(\text{rev/min})}$$

$$\therefore \quad \frac{40}{100} \times \frac{70}{75} = \frac{\text{output speed}}{675} \quad (\text{rev/min})$$

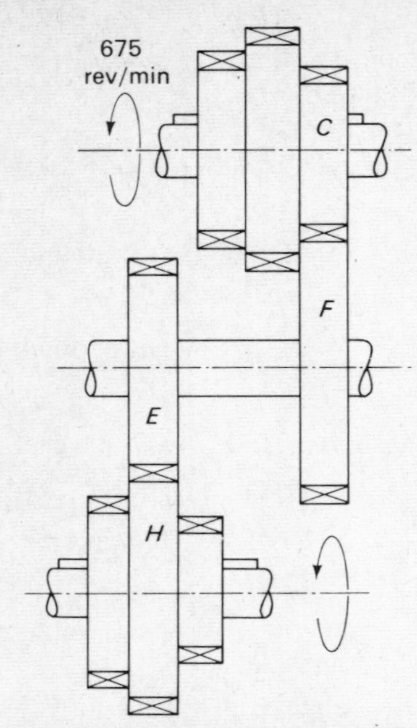

Fig. 8.11 Lowest speed arrangement

$$\therefore \quad \frac{40}{100} \times \frac{70}{75} \times 675 = \text{output speed}$$

$$= 252 \text{ rev/min}$$

$\therefore$ lowest spindle speed available = 252 rev/min.

(d) The highest speed available will employ the largest driving gears in engagement with the smallest driven gears.
Therefore

$$\frac{T_B}{T_E} \times \frac{T_F}{T_K} = \frac{\text{output speed}}{\text{input speed}} \quad \frac{(\text{rev/min})}{(\text{rev/min})}$$

$$\therefore \quad \frac{70}{70} \times \frac{100}{45} = \frac{\text{output speed}}{675} \quad (\text{rev/min})$$

$$\therefore \quad \frac{70}{70} \times \frac{100}{45} \times 675 = \text{output speed}$$

$$= 1500 \text{ rev/min}$$

$\therefore$ highest spindle speed available = 1500 rev/min.

8.3 Pulley-drive calculations

Toothed belt-drive systems are very similar to gear drives. A toothed belt acts, in effect, as an idler in a simple gear train, the input and output rotations being in the same direction. Thus toothed belt-drive calculations are identical to those for simple gear trains.

For flat- and vee-belt drives, calculations are made as shown in the following examples:

Example 1 The driving arrangement for a centre lathe headstock is shown in Fig. 8.12.

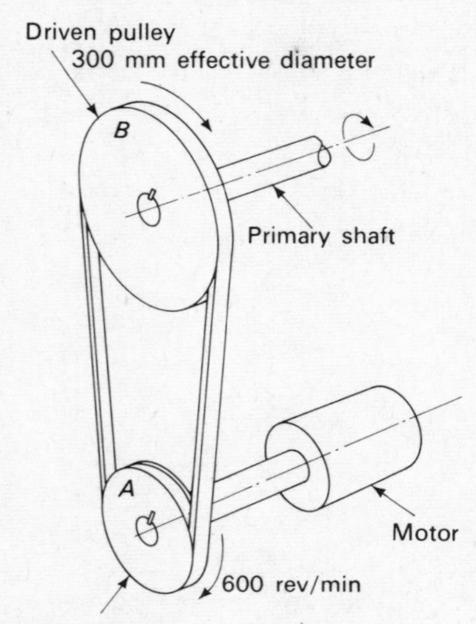

Fig. 8.12 Belt-drive system

The rotational speed of the primary shaft can be calculated as follows:

$$\frac{D_A}{D_B} = \frac{\text{speed of driven pulley}}{\text{speed of driving pulley}}$$

$$\frac{(\text{rev/min})}{(\text{rev/min})} \quad \text{or} \quad \frac{(\text{rev/s})}{(\text{rev/s})}$$

where

D_A = effective diameter of pulley A
D_B = effective diameter of pulley B.

Therefore

$$\frac{200}{300} = \frac{\text{speed of driven pulley}}{600} \quad (\text{rev/min})$$

$$\therefore \quad \frac{200 \times 600}{300} = \text{speed of driven pulley}$$

$$= 400 \text{ rev/min}$$

$\therefore$ rotation speed of primary shaft = 400 rev/min.

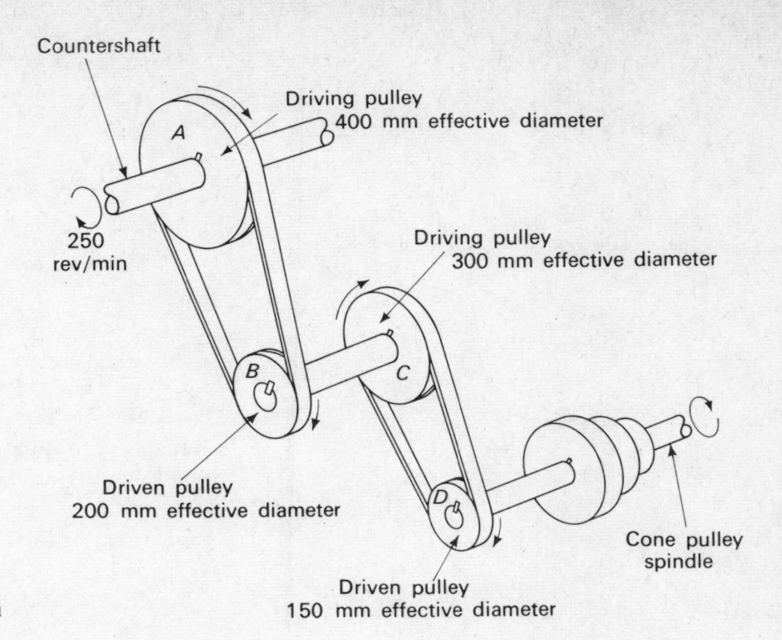

Fig. 8.13 Belt-drive system

Example 2 A certain grinding machine is driven from a countershaft by means of a system of pulleys and belts as shown in Fig. 8.13.

The rotational speed of the cone pulley spindle can be calculated as follows:

$$\frac{D_A}{D_B} \times \frac{D_C}{D_D} = \frac{\text{output speed}}{\text{input speed}} \quad \frac{(\text{rev/min})}{(\text{rev/min})}$$

$$\text{or} \quad \frac{(\text{rev/s})}{(\text{rev/s})}$$

where

D_A = effective diameter of pulley A
D_B = effective diameter of pulley B
D_C = effective diameter of pulley C
D_D = effective diameter of pulley D.

Therefore

$$\frac{400}{200} \times \frac{300}{150} = \frac{\text{output speed}}{250} \quad (\text{rev/min})$$

$$\therefore \quad \frac{400}{200} \times \frac{300}{150} \times 250 = \text{output speed}$$

$$= 1000 \text{ rev/min.}$$

$\therefore$ rotational speed of cone pulley spindle = 1000 rev/min.

8.4 Force-ratio and movement-ratio calculations

The terms *force ratio* and *movement ratio* are demonstrated in the following examples:

Example 1 The pitch of the single-start screw thread in a lifting jack is 5 mm, and the effective length of the operating handle is 210 mm. A force of 200 N, applied at the end of the handle, is required to operate against a downward force of 10 kN, as shown in Fig. 8.14. Calculate:
(a) The force ratio and
(b) the movement ratio of the lifting jack.

(a) The downward force of 10 kN is operated against by a force of only 200 N applied to the jack handle. Therefore the jack mechanism provides a useful force advantage. This is called the *force ratio* and is calculated as follows:

$$\text{force ratio} = \frac{\text{output force}}{\text{input force}}$$

$$= \frac{\text{force raised}}{\text{force applied}} = \frac{10\,000}{200} \quad \frac{(\text{N})}{(\text{N})}$$

$$= 50/1$$

$\therefore$ force ratio = 50:1.

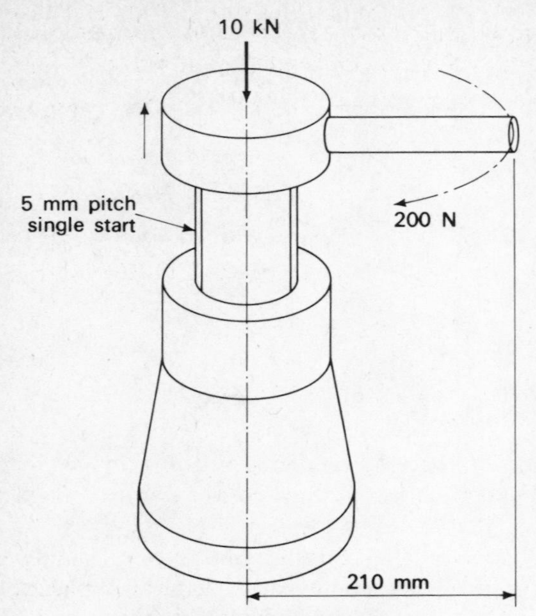

Fig. 8.14 Lifting process

(b) The force ratio was found to be very much to the advantage of the jack operator. However, this advantage must be *paid* for in some way. This repayment is called the *movement ratio* and is calculated as follows:

$$\text{movement ratio} = \frac{\text{movement of input force}}{\text{movement of output force}}.$$

Now, in one revolution of the jack handle, the input force will move around a circular path of radius 210 mm.

Therefore

$$\text{movement of input force} = 2 \times \pi \times \text{radius mm.}$$

At the same time, the force on the jack will be raised by one pitch of the screw thread.

Therefore

$$\text{movement of output force} = 5\ \text{mm.}$$

Finally then,

$$\text{movement ratio} = \frac{2 \times \pi \times 210}{5} \quad \frac{(\text{mm})}{(\text{mm})}$$

$$= \frac{2 \times 22 \times 210}{7 \times 5}$$

$$= 264/1$$

$\therefore$ <u>movement ratio = 264 : 1.</u>

Example 2 A casting is lifted from the table of a planing machine using a pulley-block system. If the casting acts downward with a force of 1200 N and the operator is required to provide a force of 60 N, calculate the force ratio of the lifting mechanism.

$$\text{Force ratio} = \frac{\text{output force}}{\text{input force}} = \frac{1200}{60} \quad \frac{(\text{N})}{(\text{N})}$$

$$= 20/1$$

$\therefore$ <u>force ratio = 20 : 1.</u>

Example 3 The knee of a horizontal milling machine produces a downward force of 1800 N. The knee is raised by the application of a force of 60 N to the knee elevating handle. Calculate the force ratio of the elevating screw mechanism.

$$\text{Force ratio} = \frac{\text{output force}}{\text{input force}} = \frac{1800}{60} \quad \frac{(\text{N})}{(\text{N})}$$

$$= 30/1$$

$\therefore$ <u>force ratio = 30 : 1.</u>

Example 4 If the net pull in a lathe spindle driving belt is 625 N and the cutting force is 4000 N, calculate the force ratio of the driving mechanism.

$$\text{Force ratio} = \frac{\text{output force}}{\text{input force}} = \frac{4000}{625} \quad (\text{N})$$

$$= 6{\cdot}4/1$$

$\therefore$ <u>force ratio = 6·4 : 1.</u>

Example 5 A gear wheel having 25 teeth drives another gear wheel having 125 teeth. Calculate their movement ratio.

$$\text{Movement ratio} = \frac{\text{movement of input force}}{\text{movement of output force}}.$$

Now, the input force must be applied to the driving gear wheel. During one revolution of the driving gear wheel, the driven gear wheel will rotate

$$\frac{25}{125} = \frac{1}{5} \quad \text{rev.}$$

$$\therefore \quad \text{movement ratio} = \frac{1}{1/5} \quad \frac{(\text{rev})}{(\text{rev})}$$

$$= 5/1$$

$\therefore$ <u>movement ratio = 5 : 1.</u>

Example 6 In Fig. 8.15, the manual traverse mechanism for the saddle of a centre lathe is shown simplified. Calculate the movement ratio of the mechanism.

$$\text{Movement ratio} = \frac{\text{movement of input force}}{\text{movement of output force}}.$$

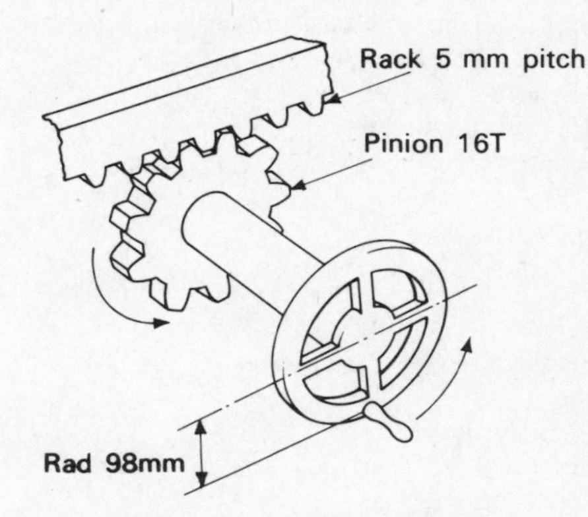

Fig. 8.15 Traverse mechanism

Now the input force must be applied at the handle. During one revolution the handle will move

$$2 \times \pi \times \text{radius (mm)} = 2 \times (22/7) \times 98 \text{ mm}.$$

At the same time the saddle will move along the rack:

$$16 \text{ teeth} = 16 \times 5 \text{ mm}.$$

Therefore

$$\text{movement ratio} = \frac{2 \times (22/7) \times 98}{16 \times 5} \quad \frac{(\text{mm})}{(\text{mm})}$$

$$= \frac{2 \times 22 \times 98}{16 \times 7 \times 5} = \frac{44 \times 14}{80}$$

$$= 7{\cdot}7/1$$

$$\therefore \quad \underline{\text{movement ratio} = 7{\cdot}7:1.}$$

Example 7 A drilling machine has a driving-belt speed of 396 m/min and drives a 20 mm diameter reamer at 30 rev/min. Calculate the movement ratio employed.

$$\text{Movement ratio} = \frac{\text{movement of input force}}{\text{movement of output force}}.$$

Now, the input force is at the driving belt which moves at

$$396 \text{ m/min} = 396 \times 1000 \text{ mm/min}.$$

The output force will be at the reamer cutting edges, whose circumference speed will be

$$30 \times \pi \times \text{diameter} = 30 \times \tfrac{22}{7} \times 20 \quad \text{mm/min}.$$

$$\therefore \quad \text{movement ratio} = \frac{396 \times 1000}{30 \times (22/7) \times 20}$$

$$= \frac{396 \times 1000 \times 7}{30 \times 22 \times 20} = \frac{210}{1}$$

$$\therefore \quad \underline{\text{movement ratio} = 210:1.}$$

Summary

Friction may be considered as a force which tends to resist sliding motion between two surfaces in contact. It may, therefore, be employed as a driving force in circumstances where sliding or slipping must be prevented. Typical applications of friction as a driving force are pulley and belt drives, friction clutches and taper-shank tool drives.

Machine-tool drive systems generally employ gears or driving belts and pulleys to transmit rotary motion. The relationship between the input speed and output speed of any gear- or belt-driven system can be calculated as explained in sections 8.2 and 8.3.

The relationship between the force applied to any mechanism and the useful force obtained from it, is called the *force ratio*. This force ratio will be of advantage to the operator of the mechanism.

To offset the advantage provided by the force ratio, the mechanism will require a considerably greater movement of the input force in comparison with the resulting output movement. The relationship between input and output movements is called the *movement ratio*. The force ratio and movement ratio of any mechanism can be calculated as explained in section 8.4.

Questions

1. Describe three machine-tool applications where friction is employed as a driving force.
2. The gearwheels in a compound gear train are:

First driver	35 teeth
First driven	50 teeth
Second driver	25 teeth
Second driven	80 teeth

(a) If the input shaft rotates at 320 rev/min, calculate the rotational speed of the output shaft, in rev/min.
(b) If the speed of the input shaft is increased to 16 rev/s, calculate the speed of the output shaft, in rev/s.

3. If the leadscrew of a lathe has a lead of 6 mm and the meshing gears in the quick-change gearbox for cutting a lead of 4 mm are 14 T driver and 27 T driven, determine the ratio of the pick-off gearing required between the lathe spindle and quick-change gearbox.
4. The spindle of a bench-mounted, sensitive drilling machine is belt driven. The cone pulleys on the motor (driving) shaft are identical to those on the machine spindle, having diameters of 80 mm, 160 mm and 240 mm respectively. If the driving shaft rotates at 3 rev/s, calculate the highest and lowest spindle speeds available.
5. For the all-geared headstock shown in Fig. 8.10 calculate:
   (a) the spindle speed for the gear arrangement shown
   (b) the highest spindle speed available, given that:
      (i) The power input shaft is driven at 700 rev/min
      (ii) The gears have the following numbers of teeth:

Gear A	45 T	B	60 T	C	35 T
Gear D	65 T	E	50 T	F	75 T
Gear G	55 T	H	70 T	K	45 T

6. In a certain lifting mechanism, an input force of 160 N enables a force of 8000 N to be raised. Calculate the force ratio involved.
7. A dividing head contains a single-start worm driving a worm wheel having 40 teeth. Calculate the movement ratio of the mechanism.
8. A cylindrical grinding machine has a driving belt speed of 5·5 m/s and drives a workpiece of 70 mm diameter at 1 rev/s. Calculate the movement ratio employed.
9. The workpiece and table of a pillar drilling machine together produce a downward force of 300 N. If the table is raised by a force of 25 N applied to a handle of 224 mm effective radius, which operates an elevating screw of 8 mm lead, calculate:
   (a) The force ratio, and
   (b) the movement ratio,
   of the mechanism.

Answers

2. (a) 70 rev/min (b) 3·5 rev/s
3. 9:7
4. 9 rev/s highest
1 rev/s lowest
5. (a) 600 rev/min
(b) 1400 rev/min
6. 50:1
7. 40:1
8. 25:1
9. (a) 12:1 (b) 176:1.

9. Materials

The craft student will already be familiar with the general properties and applications of plain carbon steels and various non-ferrous metals. He must now extend his knowledge to include properties and uses of other materials. The principles of processes which affect the properties of materials, must also be clearly understood.

This chapter examines the following topics:

9.1 Definitions of properties
9.2 Properties of materials
9.3 Pre-machining processes
9.4 Heat treatment processes.

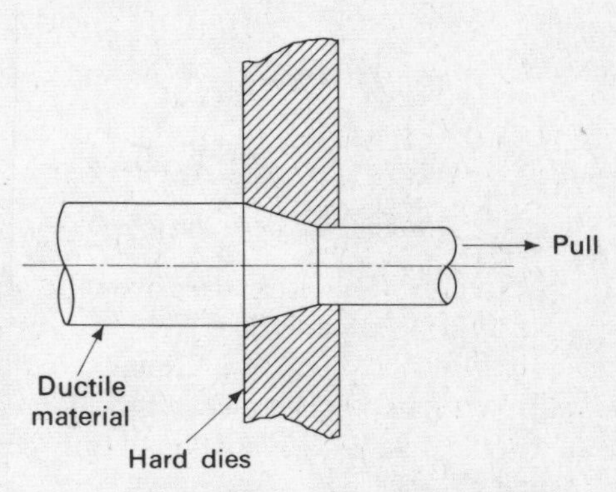

Fig. 9.1 Need for hardness and ductility

9.1 Definitions of properties

The physical properties of any material must be known, in order that it may be suitably employed. The following terms are used to indicate the physical properties of a material.

(a) DUCTILITY. This is the ability of a material to be permanently deformed without breaking, when a force is applied.

This property is generally associated with *cold drawing* processes, e.g., wire drawing.

(b) HARDNESS. This property is the resistance of a material to surface penetration, scratching or wear.

Figure 9.1 illustrates the process of wire drawing and the need for *hardness* and *ductility.*

(c) TENSILE STRENGTH. This property is the resistance of a material to tensile (stretching) forces.

(d) COMPRESSIVE STRENGTH. This is the ability of a material to withstand compressive (crushing) loads.

(e) SHEAR STRENGTH. This property is the ability of a material to resist shearing forces.

(f) TOUGHNESS AND IMPACT RESISTANCE. These properties are very similar and indicate a material's ability to withstand hard use and shock loads without breaking.

Figure 9.2 illustrates a lifting arrangement requiring high tensile strength, shear strength and toughness.

(g) MALLEABILITY. This is the ability of a material to be hammered or rolled into shape, without breaking.

Figure 9.3 illustrates a forming process requiring the properties of malleability and impact resistance.

(h) BRITTLENESS. This is the property of breaking with very little distortion.

Brittleness is often referred to as *shortness. Hot-shortness* is brittleness at high temperatures, and *cold-shortness* is brittleness at low temperatures.

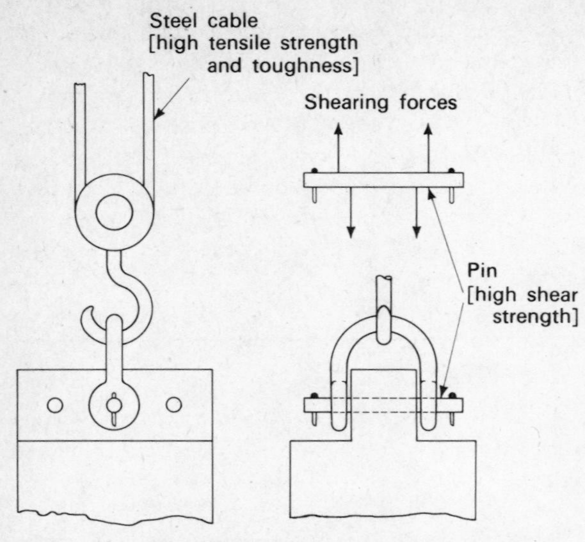

Fig. 9.2 Need for strength and toughness

(k) MACHINABILITY. This term describes the ability of a material to be machined. When discussing the machinability of a material, such factors as *tool life* and *surface finish* must be considered, in addition to the rate at which machining can be carried out.

The main physical properties which affect the machinability of a material are hardness and ductility. The greater its hardness, the lower will be a material's machinability. High ductility also reduces the machinability of a material, due to the likelihood of a poor surface finish.

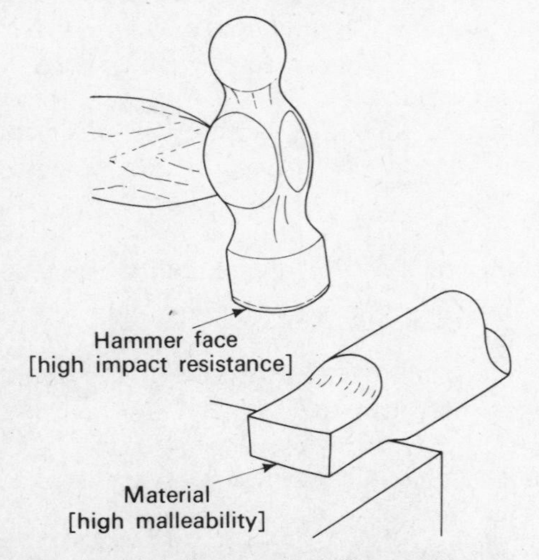

Fig. 9.3 Need for malleability and impact resistance

9.2 Properties of materials

The materials to be discussed at this time are standard alloy steels, non-ferrous alloys, and plastics.

ALLOY STEELS. Alloy steels are plain carbon steels with the addition of such elements as nickel, chromium, molybdenum, tungsten, vanadium and manganese. The effects of these elements are discussed in Vol. 2.

In general, alloy steels have the following properties:

(a) High hardness
(b) High tensile strength
(c) Good corrosion resistance
(d) Good heat resistance
(e) Good machinability (in softened condition).

The principle alloy steels used in engineering are nickel steels, chromium steels, and nickel–chromium steels.

(a) *Nickel Steels.* These are steels containing 0·4 per cent carbon, with various additions of nickel. Nickel steels have greater strength and impact resistance than equivalent plain carbon steels, and also have high resistance to corrosion and thermal expansion.

Nickel steels are used for such components as steam-turbine blades, internal-combustion-engine valves and parts of measuring instruments.

(b) *Chromium steels.* These are plain carbon steels with various additions of chromium. Chromium steels are extremely hard and have good corrosion resistance. These materials are used for such articles as drawing dies (Fig. 9.1) and ball bearings.

(c) *Nickel–chromium steels.* This is an extremely important group of alloy steels, combining the improved properties of the previous types. With suitable nickel and chromium contents, these alloy steels have tensile strengths up to five times that of plain carbon steels. For this reason nickel–chromium steels are widely used for 'high tensile' bolts and heavily loaded shafts.

An 'air-hardening' alloy steel is produced when suitable additions of nickel and chromium are made to low-carbon steel. This is particularly useful for the manufacture of components having complex shapes, which would distort if hardened by normal heat-treatment processes.

These and other alloy steels are considered in greater detail in Vol. 2.

NON-FERROUS ALLOYS. These are alloys which do not contain iron.

In general, non-ferrous alloys have the following properties:

(a) Good corrosion resistance
(b) Good machinability
(c) Good electrical conductivity
(d) High tensile strength
(e) High thermal conductivity.

The non-ferrous alloys most widely used in engineering are shown in Fig. 9.4.

(a) *Brass*. Brass is an alloy of copper and zinc, its properties depending on the amount of zinc present. Brass containing 30 per cent zinc (cartridge brass) may have a tensile strength one and a half times that of mild steel, combined with high malleability and ductility. These properties enable it to be cold worked by such processes as drawing.

Brass containing 40 per cent zinc (Muntz metal) has excellent corrosion resistance and is readily hot worked. It has poor machinability, but the addition of 3 per cent lead greatly improves this property.

Brazing Brass, containing 50 per cent zinc, is used for brazing, due to its relatively low melting temperature.

(b) *Tin bronze*. This is an alloy of copper and tin. Cast tin bronze, containing 10 per cent tin, has a tensile strength equal to that of low-carbon steel, and is widely used for bearings and bushes.

The addition of 1 per cent phosphorus to cast tin bronze results in *phosphor bronze*. This material has good wear resistance, and is used mainly for heavy duty bearings.

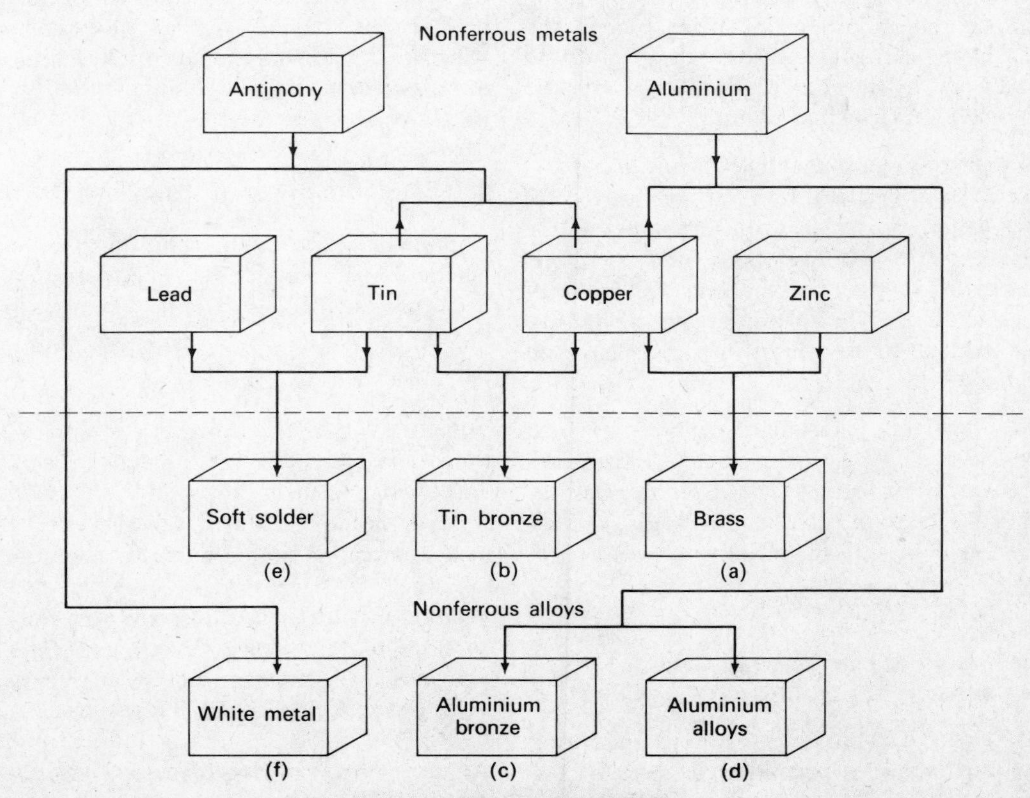

Fig. 9.4 Non-ferrous metals and alloys

(c) *Aluminium bronze*. This is an alloy of copper and aluminium and is best known for its use in casting processes.

Both sand castings and die castings are manufactured from aluminium bronze containing 10 per cent aluminium. This material has good corrosion resistance and will retain its strength at high temperatures.

(d) *Aluminium alloys*. These are alloys of aluminium and copper, only small amounts of copper being present.

Aluminium alloys are considerably lighter than steel or copper alloys, but nevertheless have good strength properties.

An aluminium alloy containing 12 per cent copper is frequently used for castings which must withstand shock loads and high stresses. Due to its good strength at high temperatures and lightness, this material is widely used for the production of engine pistons.

(e) *Soft solder*. Soft solder is an alloy of lead and tin in various proportions. It has a low melting point but poor strength properties. The student will already have learned that strength is given to soldered joints by sweating together large areas or using special joints such as seams.

(f) *White metal*. This is an alloy of tin, copper and antimony, and is used for the production of bearings which support fast rotating shafts. White metal has a low melting point, and therefore, should bearing lubrication be obstructed, the shaft will not seize in the bearing, thus avoiding extensive damage. Instead the bearing will 'run' (melt), and this is easily replaceable.

PLASTICS. Plastics are non-metallic materials which are capable of flowing, under suitable conditions, to take up a desired form. Many forms are possible, including sheets, rods, tubes, and mouldings.

In general, plastics have the following properties:

(a) Excellent corrosion resistance
(b) Good machinability
(c) Good electrical resistance (Insulation)
(d) Low strength.

The plastics most commonly used for engineering purposes are polythene, perspex, nylon, P.V.C., and rubber.

(a) *Polythene*. This material has low strength, but offers high resistance to the flow of electricity. It is widely used therefore, as an electrical insulator, e.g., for cable coverings. It is also used for flexible containers, these being manufactured by moulding processes.

(b) *Perspex*. This is a transparent material with good impact resistance. It is widely used for the manufacture of eye shields fitted to machine tools.

(c) *Nylon*. Due to its relatively high strength, low frictional properties and good wear resistance, this material is often used for gears and bearings. Under such circumstances, lubrication is not a necessity.

(d) *P.V.C.* This material has wide-ranging properties, including high electrical resistance. It is available in many forms, varying from highly flexible to very rigid. It is used for the manufacture of plastic clothing, and moulded components such as car-battery cases.

(e) *Rubber*. This material has good vibration-damping properties, and can be extremely tough and flexible. Its high-friction properties make it very suitable for the manufacture of items such as driving belts.

9.3 Pre-machining processes

Components are usually finished to shape by machining processes. The material supplied for machining will be of suitable form, produced by casting, 'hot-working' processes or 'cold-working' processes.

CASTING. The casting process consists of filling a shaped mould with molten metal, which solidifies to the required form on cooling.

Sand moulds are used for the production of *sand castings*, a simple example being shown in Fig. 9.5.

Metal moulds, called dies, are used for the production of *die castings*, as shown simplified in Fig. 9.6. The die-casting process is mainly restricted to non-ferrous metals and alloys.

HOT-WORKING PROCESSES. These are processes where the material is forced into shape at high tempera-

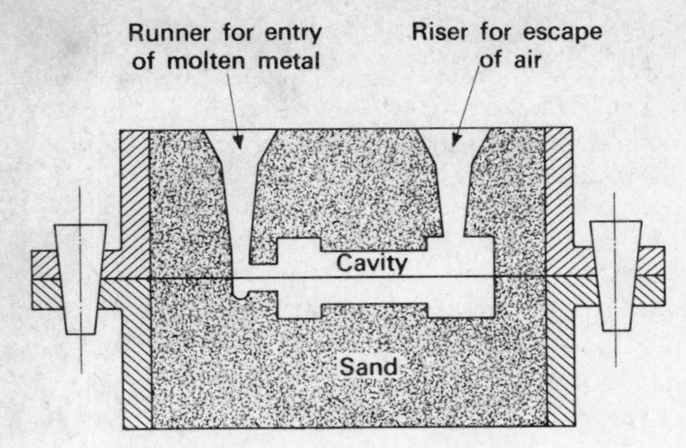

Fig. 9.5 Sand casting

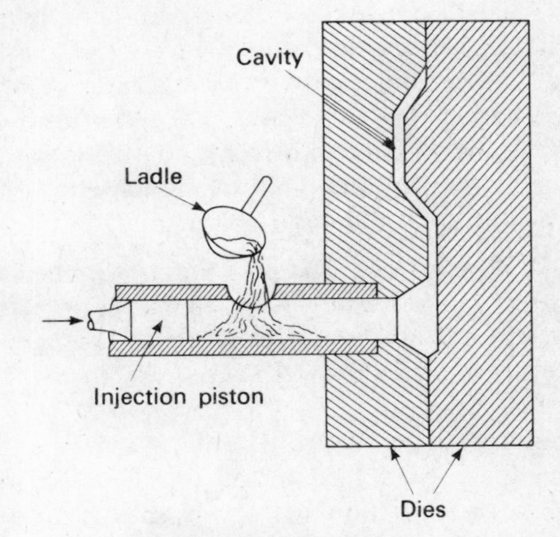

Fig. 9.6 Die casting

tures. The principal hot-working processes are forging and hot rolling.

(a) *Forging.* Forging consists of shaping hot material using a hammer or press. The use of hand-forging techniques is restricted to small components, produced in limited quantities. For large forgings and quantity production, power hammers or hydraulic presses are employed.

Metal dies may be fitted to the hammer and anvil for forging large quantities of identical components. When these are used, the process is called *drop forging.*

(b) *Hot rolling.* Hot metal ingots may be reduced in cross-section to produce metal plates, by passing them between plain steel rolls. Similarly round, square and hexagon bars, or angles, tees and channel sections, may be formed, using shaped rolls in several stages. The principle of rolling is illustrated in Fig. 9.7.

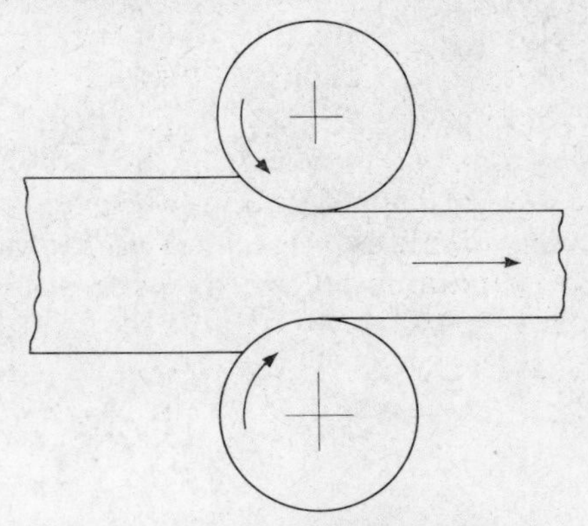
Fig. 9.7 Principle of rolling

COLD-WORKING PROCESSES. These are processes where the material is forced into shape at room-temperatures. Cold-working processes are generally applied to the finishing stages of production, in order to gain such benefits as the following:

(a) The production of a bright, smooth finish
(b) The production of components to accurate dimensions.

The principal cold-working processes are cold rolling and drawing.

(a) *Cold rolling.* Strips and sheets are finished by cold rolling. A *four-high mill* is used as shown in Fig. 9.8, which illustrates the production of thin

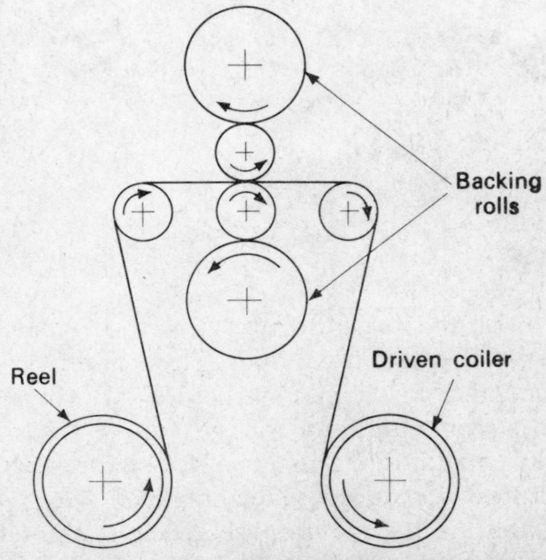

Fig. 9.8 Production of thin strip steel

strip steel. The large backing rolls support the working rolls, to maintain uniform strip thickness.

(b) *Drawing*. This is a cold-working process which relies entirely on the ductility of the material being drawn. Both solid sections and tubes are produced by drawing material through dies, and all wire is manufactured by this process, as shown in Fig. 9.1.

'Cup-shaped' components may be manufactured by a cold-working process called *deep drawing*. A *blank* of material is pressed to shape in a die, as shown in Fig. 9.9.

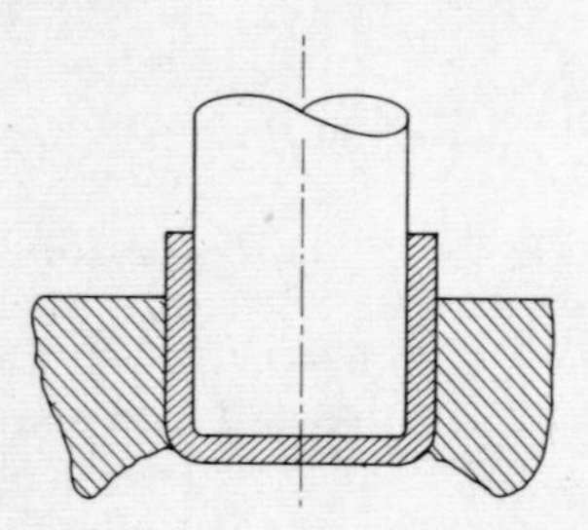

Fig. 9.9 Deep drawing

PROPERTIES RESULTING FROM PRE-MACHINING PROCESSES. All metals and alloys are *crystalline* in structure. Figure 9.10 illustrates the crystalline structure of pure iron when magnified under a microscope.

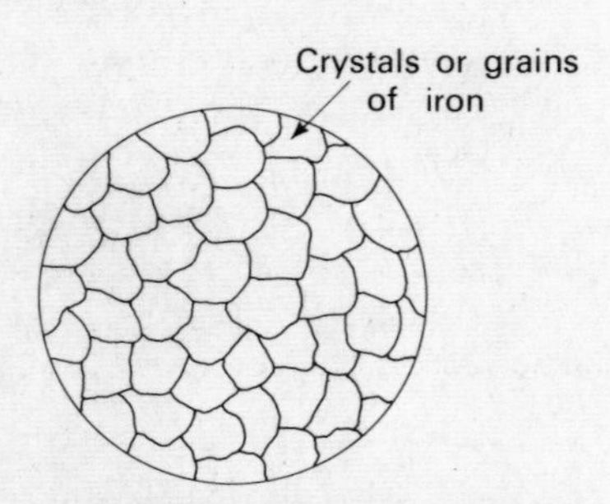

Fig. 9.10 Structure of pure iron

The size and formation of crystals in any material have considerable effect on its physical properties. In general, a coarse grain structure (large crystals) lacks strength and toughness, and tends to be brittle. In addition, a major effect of coarse grain structures is reduced machinability.

Coarse grain structures are generally the result of holding materials at high temperatures, for lengthy periods, without suitable subsequent heat treatment.

The formation of crystals in a material is mainly dependent upon the processes to which it has been subjected.

Casting effects. Casting is a very precise process, requiring considerable skill and experience. If gases are trapped in the molten metal during casting, 'blow holes' may result in the structure. These may not appear until machining is carried out, whereupon the casting may have to be scrapped.

All metals contain impurities which tend to settle in the portion of a casting which solidifies last. This effect causes brittleness in such regions.

Sand castings tend to have coarse grain structures, which reduce physical properties, as mentioned earlier. Die castings do not suffer from this problem, having much finer grain structures. This is due to the rapid cooling of the material upon contacting the metal dies.

Hot-working effects. The cast ingots supplied for hot-rolling and forging have coarse grain structures. The heavy working of the metal during these processes breaks down the structure to a much finer form.

During forging, not only is the grain structure of the material refined (made finer), but it also takes up a formation with a definite grain-flow direction. This effect is illustrated in Fig. 9.11 which shows the internal structure of a forged hub. Note that the grain flow is such that no weaknesses occur due to sudden changes of direction.

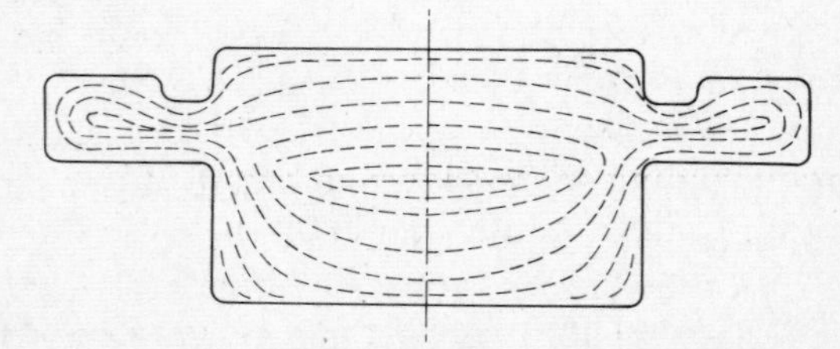

Fig. 9.11 Internal structure of forged hub

Cold-working effects. The general effects of cold working a material are increased hardness, tensile strength and brittleness, and reduced ductility.

The increased hardness due to cold-working processes is called *work hardening*. This effect occurs in many workshop processes, including shearing and bending, but is most noticeable in severe cold-working processes such as cold rolling and drawing.

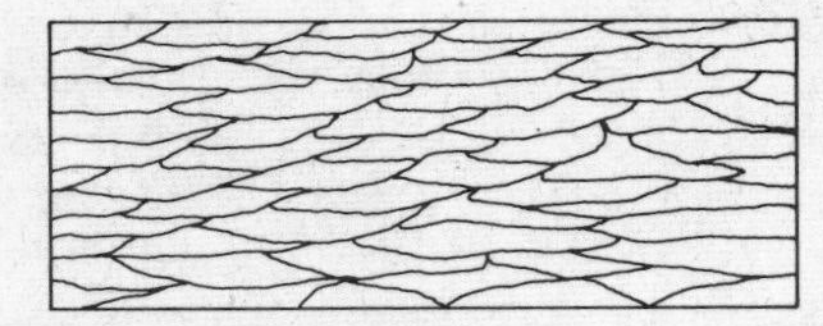

Fig. 9.12 Structural effect of cold rolling and drawing

Figure 9.12 shows the structural effect of processes such as cold rolling and drawing. Note that the grains are flattened, and elongated in the direction of rolling or drawing. During such processes, it is often necessary to remove work hardening, by suitable heat treatment, before cold working can be completed.

9.4 Heat-treatment processes

Heat treatment is a means of improving the physical properties of a material, by subjecting it to controlled heating and cooling. The heat treatment of non-ferrous alloys is considered in Vol. 2. For the present, the heat treatment of steels only will be considered.

When steel is heated, it undergoes structural changes. It is not necessary to examine the precise nature of these changes at this time, but the temperature range during which structural changes occur is of considerable interest.

When, for example, a medium carbon steel component is heated, its internal structure begins to change at 700°C and is completely changed to a new form at a temperature of about 800°C. For this steel 700°C is said to be its *lower critical point* and 800°C its *upper critical point.*

The critical points of any steel must be known, in order that heat-treatment processes may be carried out satisfactorily. The diagram shown in Fig. 9.13 indicates the critical points of all plain carbon steels.

The following points should be noted:

(a) For all plain carbon steels, the lower critical point is at 700°C.

(b) The upper critical point varies with the carbon content of the steel

(c) 0·87 per cent carbon steel has only one critical point, this being at 700°C.

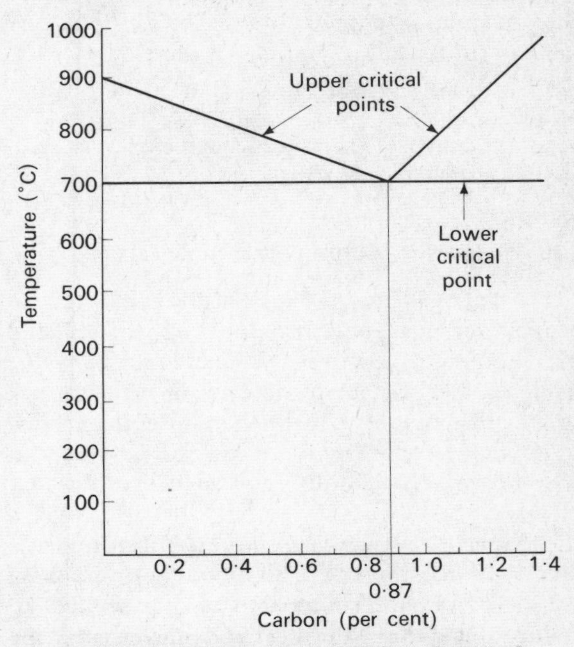

Fig. 9.13 Critical points of plain carbon steels

The principal heat treatments given to steels are *annealing*, *normalizing*, *hardening*, and *tempering*.

(a) ANNEALING. This process is carried out to make steel as soft as possible. This will make the steel more suitable for cold working and machining.

Annealing is carried out by heating the steel to a temperature just above its upper critical point, allowing it to soak at this temperature, and then cooling it very slowly. The speed of cooling is most important and can best be controlled by turning off the furnace and allowing the steel to cool down within it.

(b) NORMALIZING. This process is carried out to refine the grain structure of a steel. Normalizing may be necessary to counteract such effects as the following:

(i) Coarse grain structures, due to lengthy processes at high temperatures, e.g., forging

(ii) Distorted grain structures, due to such processes as hammering and cold rolling
(iii) Work-hardening effects, due to cold working.

Normalizing is carried out by heating the steel to a temperature just above its upper critical point, and then allowing it to cool in still air. This is very similar to annealing, but the cooling rate is less slow.

(c) HARDENING. This process increases the hardness of a steel in order that it may:

(i) Resist wear, or
(ii) be capable of cutting other materials.

Hardening is carried out by heating the steel to a temperature just above its upper critical point, and then cooling it very rapidly. Fast cooling may be achieved by quenching the steel in water or oil. Oil quenching is less drastic than water quenching, and is often preferred for articles having thin sections which might distort if quenched in water.

(d) TEMPERING. A hardened steel will have a brittle structure. This may lead to chipping or cracking of a hardened steel component when in service. To reduce this danger, the steel is tempered after the hardening process. Tempering causes a reduction in hardness, but considerably improves the toughness and impact resistance of a hardened steel.

Tempering is carried out by re-heating the steel to a temperature well below its lower critical point, and then cooling it, either in air or by quenching. The precise temperature required for tempering is dependent upon the type of hardened component. This temperature is often judged by the colour of an oxide film which forms on the freshly polished surface of a heated steel article. Tempering temperatures used for typical workshop articles are listed in Table 9.1.

Table 9.1

Article	Temperature °C	Temper Colour
Scrapers	230	Pale straw
Turning tools	240	Dark straw
Taps, punches and reamers	250	Light brown
Press tools	260	Brown
Chisels	270	Purple
Hammer heads	280	Dark purple
Springs	300	Blue

The heat-treatment processes of hardening and tempering are usually applied to high-carbon steels and have little effect on steels with low carbon content.

CASE HARDENING. Mild steel, having a low carbon content, cannot be hardened to any useful extent by heating and quenching. If however, the carbon content is increased at the surface of a mild-steel component, this outer layer or *case* may be hardened by heating and quenching. This process, known as *case hardening*, is therefore carried out in separate stages:

(a) The production of a high carbon content outer layer (carburizing).
(b) Heat treatment.

These stages are carried out as follows:

(a) *Carburizing.* The mild-steel component is packed in an airtight iron box containing a substance rich in carbon, e.g., charcoal. The box and its contents are then heated in a furnace at a temperature of about 900°C for a period of about six hours. During this time, carbon will penetrate the surface of the mild-steel component, forming a high-carbon steel outer layer approximately 1 mm deep.

(b) *Heat treatment.* Having allowed the box to cool, the component is removed and heat treated in two stages.

Stage 1. The mild-steel core of the component will have a coarse grain structure due to the period spent at high furnace temperature. This structure will be weak and is, therefore, toughened by *core refining.* This consists of re-heating the component to just above the upper critical point of the core, and then quenching the component in oil.

Reference to Fig. 9.13 will indicate that the upper critical point of the mild-steel core (0·25 per cent C) is at about 870°C.

Stage 2. The high-carbon steel case is now hardened by re-heating the component to just above the upper critical point of the case, and then quenching it in water.

Note that the upper critical point of the high-carbon steel case (0·9 per cent C) is at about 780°C.

ALTERNATIVE METHOD OF CASE-HARDENING. When only a shallow depth of case is necessary, carburizing may be achieved by suspending the component in a bath of liquid cyanide. This process is carried out at a temperature of about 900°C, and in one hour a case depth of 0·2 mm may be obtained. After a period of up to two hours, the component may be removed and quenched immediately, as the core will not need refining.

Summary

The *physical properties* of a material must be known in order that it may be suitably employed. By suitable choice of materials, properties exactly fitting the requirements of a component can be made available.

The *alloying* of metals produces materials which have certain improved properties, both ferrous and non-ferrous alloys being widely used for engineering purposes. The use of *plastics* is ever increasing and although these materials will never completely replace metals, they will certainly find many more applications than at present.

The properties of materials may be altered during pre-machining processes. Such effects as *work-hardening*, during cold working, and grain growth, during hot working, considerably alter the properties of a material. These effects may be counteracted by further processes known as *heat treatments.*

Heat-treatment processes are used to improve the properties of a material. These processes consist of heating materials to certain temperatures, and then cooling them at suitable rates. The principal heat treatment processes are *annealing, normalizing, hardening* and *tempering.*

Materials such as low-carbon steels do not respond to hardening by heat treatment. However, *case-hardening* may be achieved by carburizing a component surface, refining its core, and finally hardening its outer layer or case.

The temperatures between which structural changes occur in a material, are called *critical points.* These must be known in order that heat-treatment processes may be carried out satisfactorily.

Questions

1. Name and define four properties of materials. In each case describe a workshop example where the property is desirable.
2. What are the main properties and uses of the following materials:
   (a) Alloy steels
   (b) Non-ferrous alloys
   (c) Plastics.
3. Name and describe three materials used for bearings carrying rotating shafts.
4. State suitable materials for the manufacture of each of the following:
   (a) Drawing dies
   (b) Heavily-loaded bolts
   (c) Engine pistons.
   Give reasons for each choice.
5. Name and describe three pre-machining processes.
6. Briefly compare the internal structures of sand castings and die castings.
7. Describe:
   (a) The internal structure of a cold-rolled article
   (b) The effect of work-hardening.
8. Name and describe four heat-treatment processes, with reasons for their being carried out.
9. By reference to Fig. 9.13, determine suitable heat treatment temperatures for:
   (a) Annealing 0·3 per cent carbon steel
   (b) Normalizing 0·5 per cent carbon steel
   (c) Hardening 1·0 per cent carbon steel.
10. Describe a case-hardening process and describe the properties of the case and core of a case-hardened mild-steel shaft.

10. Studies associated with materials

This chapter examines the following topics associated with materials:

10.1 Stress and strain

Most engineering components are required to carry loads of some description. These loads may tend to stretch, crush or shear a component in service. When, for example, a nut and bolt are used to secure an assembly, the following conditions occur:

(a) The bolt tends to *stretch* when the nut is tightened
(b) The assembly tends to be *crushed* between the bolt head and nut
(c) The screw threads tend to shear.

The materials chosen for the manufacture of such components must therefore have sufficient *strength* to withstand these applied loads. In section 9.1 it was stated that the properties of a material include tensile strength, compressive strength, and shear strength, these being required to resist tensile (stretching), compressive (crushing), and shearing loads respectively.

STRESS. When a material is subjected to loads it is said to be *stressed*. The amount of *stress* within a material may well cause deformation (change of shape) or even fracture to occur. These effects are produced when the stress overcomes the material's strength.

The amount of stress in a component is measured in N/m^2 and therefore depends upon the size of:

(a) The load acting on the component (N)
(b) The area of the component under load (m^2).

Thus, if two bolts, of 10 mm diameter and 20 mm diameter respectively, carry identical tensile loads, the stress in the larger bolt will be much less than that within the smaller one. This is because the load is spread over a larger area in the case of the 20 mm diameter bolt. Any increase in load on either bolt, however, will result in increased stress.

Conditions producing various types of stress are shown in Fig. 10.1.

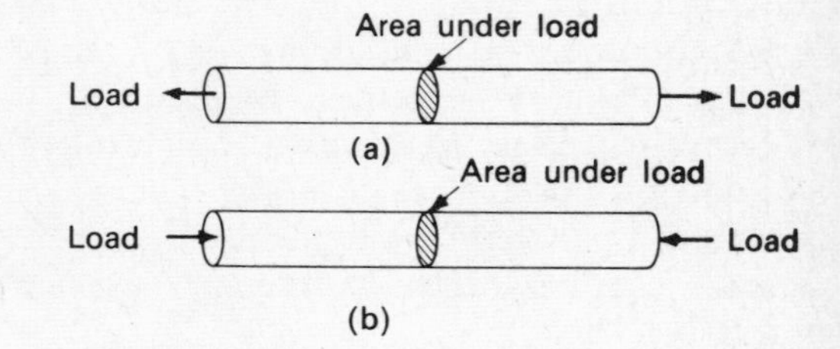

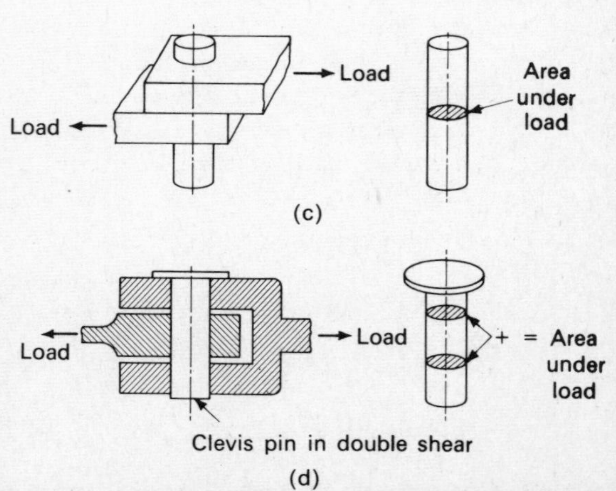

Fig. 10.1 Conditions of stress: (a) tensile; (b) compressive; (c) and (d) shear

STRAIN. When a component is stressed, it may well undergo a change in length. A bolt under load, for example, might be found to stretch from an original length of, say, 50 mm to a final length of 52 mm. In this stretched condition the bolt is said to be *strained.* The strain may be permanent or alternatively may disappear when the load on the bolt is removed. Strain may similarly occur in components subjected to compressive and shear stresses.

STANDARD TESTS. It is necessary for the engineer to know how any material will behave when it is subjected to applied loads. It is found, for example, that cast iron can withstand very large compressive loads, thus indicating its high compressive strength. Its tensile and shear strengths, however, are very low, and therefore cast iron is unsuitable for the manufacture of such items as bolts and clevis pins. Furthermore some materials will undergo considerable strain before fracture occurs due to an applied load. A material which behaves in this manner under the action of a *tensile* load is said to be *ductile*, and is therefore suitable for cold-working processes such as wire drawing.

Various types of testing machine are used to investigate the behaviour of materials under load. These machines are capable of applying varying tensile, compressive or shearing loads to material specimens. After each increase in load any change in the specimen's length is recorded and this information is used to indicate the material's main properties and capabilities.

TENSILE TESTS. These tests provide much useful information concerning a material's properties. A test piece or specimen of material is clamped between two grips, fitted to the testing machine, as shown in Fig. 10.2.

A tensile load is applied to the test piece and any extension (stretch) is recorded using a special measuring device. The load is then increased in gradual steps, extension of the test piece being measured at each increase. This procedure is repeated until the test piece finally fractures.

From the results obtained during such a test it is then possible to plot a graph showing the relationship between load and extension. A typical graph produced from the results obtained during a tensile test on a mild-steel specimen is shown in Fig. 10.3.

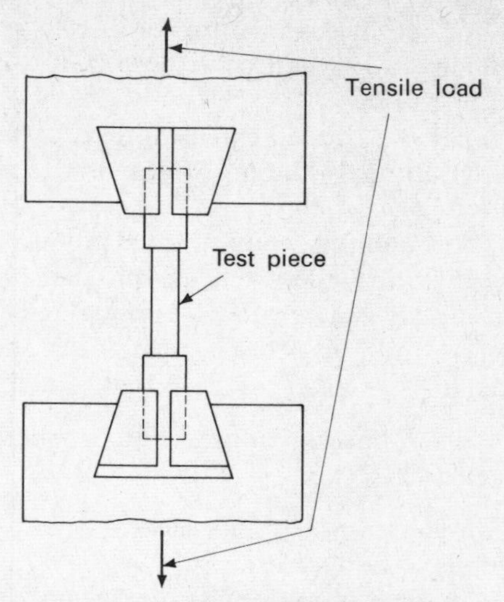

Fig. 10.2 Test piece gripping arrangement

The following points should be noted from the graph:

(a) Up to point A each increase in load produces a similar or proportional increase in extension
(b) Up to point A the material is *elastic*, i.e., the test piece would return to its original length if the load were removed. Components such as

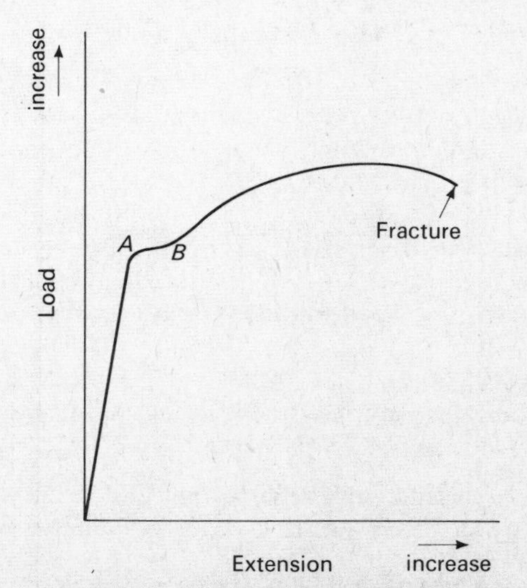

Fig. 10.3 Load–extension graph for mild steel

bolts should not, therefore, be loaded beyond this point, if permanent deformation or strain is to be avoided

(c) At point *B* a sudden extension occurs without any additional loading. This is called the material's *yield point*.

(d) Further loading beyond point *B* produces considerable permanent extension, until the test piece finally fractures. This considerable extension beyond the yield point indicates that the material is *ductile*.

Thus the following information is obtainable from such a graph:

(a) The maximum load possible if permanent strain is to be avoided
(b) The material's suitability for processes requiring ductility, e.g., wire drawing
(c) The load required to produce ductile extension.

Load–extension graphs for carbon-steel specimens after various heat treatments are shown in Fig. 10.4.

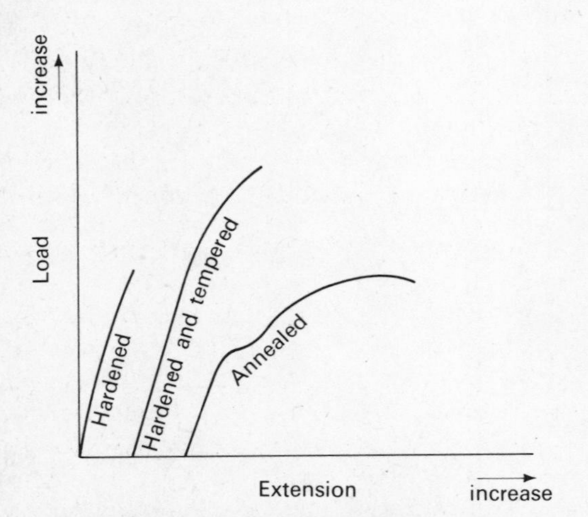

Fig. 10.4 Load–extension graphs for carbon steel

From these graphs the following points should be noted:

(a) After hardening or hardening and tempering, carbon steels do not show a definite yield point
(b) Hardened steels lack ductility, even when tempered
(c) Hardened and tempered steels are capable of withstanding very high tensile loads before fracture occurs.

Load–extension graphs for many non-ferrous metals and alloys indicate considerable ductility but no definite yield point, as shown in Fig. 10.5.

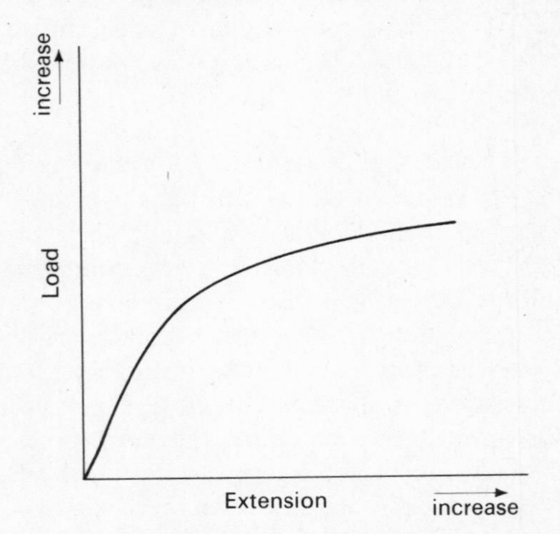

Fig. 10.5 Load–extension graph for non-ferrous metal or alloy

COMPRESSION AND SHEAR TESTS. Compression tests are conducted similarly to tensile tests, the load however being applied to 'crush' the test piece. The main information obtained from such a test is the load at which the specimen ruptures.

Shear tests are very similar to tensile tests and indicate the load at which a test pin will shear. The method of gripping a pin for testing in double shear is shown in Fig. 10.6.

10.2 Material specifications on drawings

An engineering drawing is used to provide the craftsman with all necessary information required for the manufacture of a workpiece. It will, for example, indicate the finished shape of a component, together with its dimensions. It is also necessary for information concerning the workpiece material to be specified on a drawing. This information must include the following:

(a) Material: mild steel, brass, etc.
(b) Form: black bar, drop forging, etc.

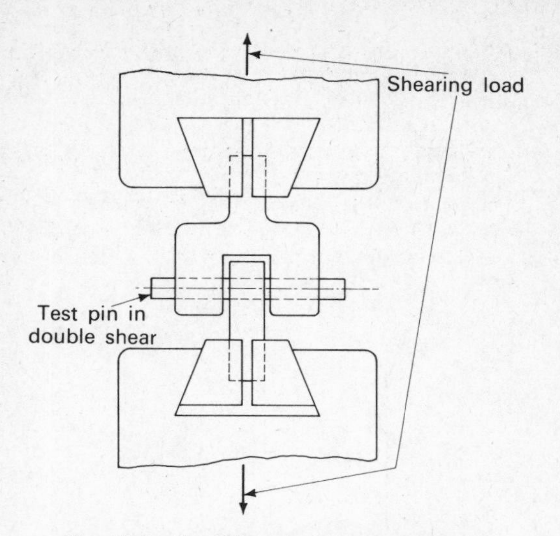

Fig. 10.6 Test pin gripping arrangement

(c) Protective finish: grease protected, cadmium plated, etc.
(d) Treatment: any required heat treatment, e.g., hardening and tempering.

Various methods are employed for showing such information on an engineering drawing, each engineering firm having its own particular style. When a drawing is concerned with only one manufactured item, the above information is generally included within the title block. A typical layout of such a title block is shown in Fig. 10.7.

Where several items or parts are shown on the same drawing sheet it is usual for a list of parts and materials to be placed close to the title block.

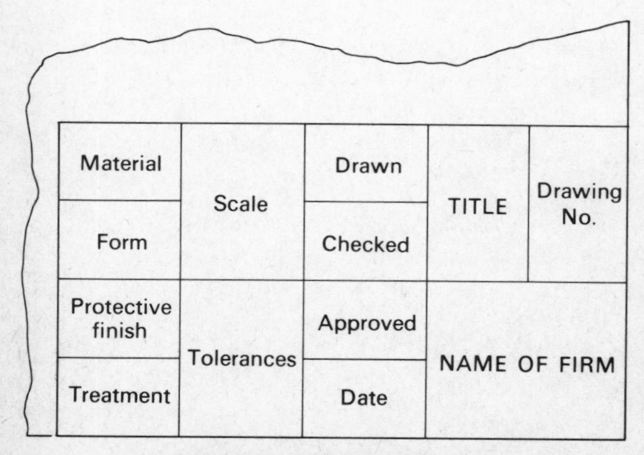

Fig. 10.7 Title block layout

Numerous types of carbon and alloy steels are used for the manufacture of engineering components. These may be supplied in many forms, the most common being bars, billets, forgings, plate, sheet, and strip. Steels supplied in these forms have been listed in British Standards 970 and 1449, where the precise contents and nature of each steel is specified. Each type of steel is designated by the letters 'En' followed by a number. For example, En 1A is a free-cutting steel for machining, containing approximately 0·1 per cent carbon. Thus, it is common practice for steels to be specified on drawings by their 'En series' number; for example,

Material: Steel—BS 970 En 16T.

Summary

The properties of any material include *tensile strength*, *compressive strength*, and *shear strength*, these being required to resist tensile, compressive, and shearing loads, respectively. When a material is subjected to such loads, it is said to be *stressed*. The stress within a material must not exceed the material's strength if permanent deformation is to be avoided. If deformation should occur then the material is said to be *strained*. This *strain* may be permanent, or alternatively may disappear when the applied load is removed. Permanent strain is the object of certain processes, e.g., drawing, and for such applications materials should be employed which are capable of being considerably strained without breaking (ductile materials).

Standard tests are carried out to establish such properties of a material as strength and ductility. During these tests, increasing loads are applied to a test piece of material until it breaks. The load at which fracture starts to occur will indicate the material's strength. Tensile tests are particularly useful, as by recording the extension of the test piece at each increase of load, a graph may be plotted to illustrate the material's behaviour. By recognizing the form of such graphs it is possible to appreciate the main properties of any material.

An engineering drawing should include all information necessary for the production of a workpiece or component. This should include material, form in which supplied, finish and any heat treatment required. Such information is usually laid out in the title block or in a separate materials list.

Questions

1. What do you understand by the strength of a material?
2. What is stress?
3. Two components are shown standing on a concrete floor in Fig. 10.8. Estimate which component causes the greater compressive stress within the concrete.

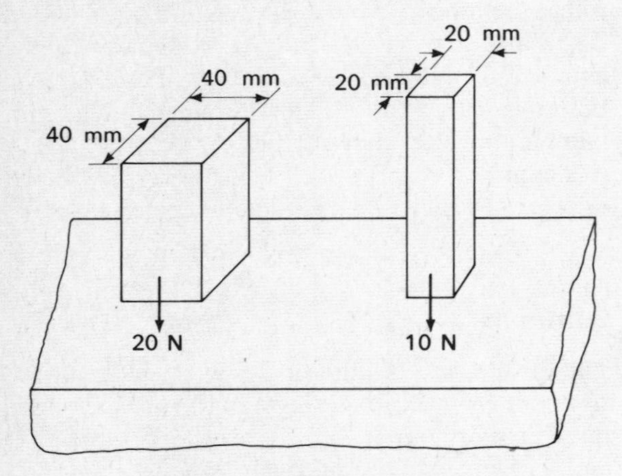

Fig. 10.8 Compressive loads

4. What is strain?
5. Describe a workshop example of (a) temporary, and (b) permanent strain.
6. Describe a standard test used to investigate the behaviour of a material under load.
7. (a) Sketch the form of graph obtained from a tensile test on a mild-steel test piece.
   (b) Explain the main information indicated by the form of this graph.
8. Load–extension graphs resulting from tensile tests on three different materials are shown in Fig. 10.9. Compare the main properties of these materials.

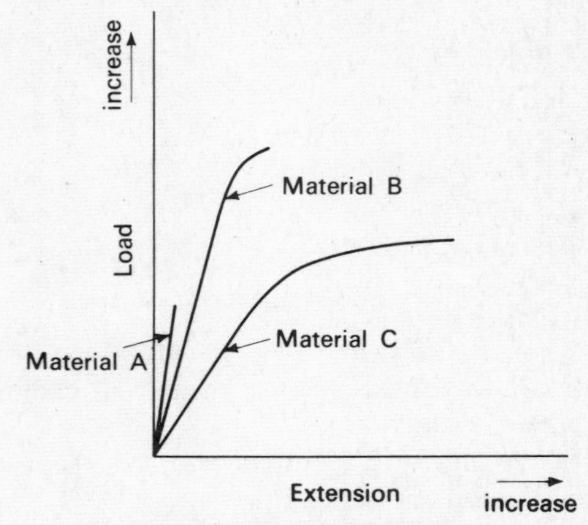

Fig. 10.9 Load–extension graphs

9. What information is required on an engineering drawing, concerning the material used for manufacture?

11. Material removal

Workpieces and components are commonly manufactured to size by the removal of unwanted material. This may entail the use of various cutting tools and processes, many of which the student will already have encountered.

This chapter examines the following topics:

11.1 Cutting tool materials
11.2 Cutting tool geometry
11.3 Throw-away tips
11.4 Types of file
11.5 Hand scraping
11.6 Lapping.

11.1 Cutting-tool materials

The student should already be familiar with high-carbon steel and high-speed steel cutting tools. However, brief revision will be given to these materials before proceeding to other types.

(a) HIGH-CARBON STEEL. This material contains between 0·8 per cent and 1·4 per cent carbon and has the following properties and features:

(i) Good toughness
(ii) Good hardness up to 250°C
(iii) Relatively inexpensive.

Typical high-carbon steel cutting tools are:

Chisels	0·8 per cent–0·9 per cent carbon
Files and taps	0·9 per cent–1·1 per cent carbon
Scrapers	1·1 per cent–1·4 per cent carbon

In addition, high-carbon steel is often employed for form tools, because of its low cost.

(b) HIGH-SPEED STEEL. This material contains tungsten, chromium and vanadium in various amounts. A popular composition is known as *18–4–1*, this containing 18 per cent tungsten, 4 per cent chromium and 1 per cent vanadium. High-speed steel has the following properties and features:

(i) Reasonable toughness
(ii) Good hardness up to 600°C
(iii) Fairly expensive
(iv) Suitable for cutting all materials, except the hardest steels
(v) Suitable for interrupted cuts, e.g., turning square section bar down to round.

This material is used mainly for drills, turning tools and milling cutters.

OTHER CUTTING-TOOL MATERIALS. High-carbon steel and high-speed steel have certain limitations to their uses as cutting-tool materials. They are, for example, unsuitable for cutting hard alloy steels, and have limited capabilities of retaining their hardness at high temperatures (red hardness or hot hardness). Cutting-tool materials with improved properties, such as red hardness, include:

(a) Super high-speed steel
(b) Stellite
(c) Cemented carbides
(d) Ceramics
(e) Diamonds.

(a) *Super high-speed steel.* This material is of similar composition to 18–4–1 high-speed steel but has an additional content of between 3 per cent and 12 per cent cobalt.

Super high-speed steel tools are hard, have good wear resistance, and reasonable toughness. These tools are widely used for machining alloy steels, particularly on capstan and turret lathes.

To offset the relatively high cost of the material, super high-speed steel tips are often butt-welded to carbon steel shanks. Very briefly, this process consists of clamping the tips and shanks together, and passing a heavy electric current between them. This causes intense heat to develop at the joint, and, with a steady pressure applied, the tip becomes firmly welded to the shank.

(b) *Stellite.* This is a cast, non-ferrous alloy containing 50 per cent cobalt together with chromium and tungsten. Stellite has poorer impact resistance than high-speed steel but may be used for interrupted cutting if rigidly supported. The main advantage of stellite as a cutting-tool material is its ability to retain extreme hardness up to temperatures in the region of 800°C.

Turning tools using stellite are of the tipped type, the tip usually being silver-soldered to a medium-carbon steel shank, using an oxy-acetylene torch. Alternatively, stellite may be deposited on to a tool shank by welding processes.

Stellite-tipped tools are widely used for cutting cast iron at high speeds, and for machining hard steels.

(c) *Cemented carbides.* There are various types of these materials, the most common being compounds of tungsten or titanium, carbon and cobalt. Minutely ground particles of tungsten or titanium with carbon are cemented together, under heat and pressure, with a softer material, such as cobalt, which provides a bond.

Cemented carbides are capable of machining at cutting speeds up to four times those possible with high-speed steel, but finer feed rates must be used. These materials have extremely good wear resistance, combined with excellent red hardness. Their brittleness and poor impact resistance however, make them unsuitable for interrupted cuts, and indeed every precaution must be taken to prevent vibrations during machining, as these can cause tool fracture.

Tungsten carbide. This tool material is mainly used for cutting cast iron, non-ferrous alloys and plastics. It is not suitable for machining steel, as the chips tend to build up, and under the heat and pressure of the cut, weld themselves to the tip. When the chips are torn away as machining proceeds, the tool tip is liable to damage and the workpiece to scoring. This damage to the tool tip is known as *cratering.*

Titanium carbide. This also is a very hard material, but has relatively good resistance to shock loads and cratering. Due to the latter, it is used mainly for machining steels at high speeds.

Mixed carbides. These are usually mixtures of tungsten and titanium carbides which combine the better properties of the previous types.

All cemented carbides are very expensive and are, therefore, supplied in the form of tips which are brazed to medium-carbon steel shanks.

(d) *Ceramics.* The term *ceramics* was originally used to describe materials manufactured from clay, but now has wider applications. One ceramic material, containing 95 per cent aluminium oxide, provides an extremely hard cutting-tool material. This has excellent red hardness, and may have a *tool life*, between regrinds, over twice that of cemented carbides.

Ceramics will cut most metals at speeds up to three times those used with high-speed steel and do not suffer from cratering effects. They are, however, extremely brittle, and vibrations must be avoided during machining. Ceramic tips cannot be satisfactorily brazed to carbon-steel shanks, and *tool bits* are therefore clamped in suitable holders.

(e) *Diamonds.* Diamond is pure carbon, and provides the hardest cutting material known. *Industrial diamonds* are cut and lapped (section 11.6) to a suitable shape, and secured in holders. Diamond-tipped cutting tools cannot withstand heavy cutting pressures or vibrations, and can only be employed for very light finishing cuts. Extremely long tool life may, however, justify their use for finishing operations, provided that suitably vibration-free machine tools are available. Diamond-tipped cutting tools are used mainly for machining non-ferrous metals and the harder plastics. They are used, for example, in the motor-car industry for finish-turning aluminium pistons and for boring connecting-rod and main bearings. The use of diamonds for machining irons and steels is not justified, as they cannot compete with carbides.

MOUNTING OF CUTTING TOOLS. As the harder grades of cutting tools cannot withstand vibrations, the student should bear in mind the basic principles of tool mounting to prevent vibration and chatter. Such features as the following should constantly be guarded against:

(a) the use of numerous and varied packing pieces beneath turning tools
(b) the mounting of tools with excessive overhang from the tool post
(c) the incorrect grinding and setting of cutting tools.

11.2 Cutting-tool geometry

The student will already appreciate that cutting tools require suitable *rake* and *clearance* angles. The rake angle is responsible for providing an efficient cutting action, and the clearance angle for preventing rubbing between the tool and workpiece.

The clearance angles applied to turning tools are generally between 6° and 10°. The rake angles vary considerably, however, depending on the tool and workpiece materials. Suitable rake angles for various tool and workpiece materials are given in Table 11.1.

Table 11.1 Cutting-tool rake angles (degrees)

Material being cut	Tool material			
	H.S.S.	S.H.S.S.	Stellite	Carbides
Mild steel	20	20	20	0–8
Grey cast iron	0–10	0–8	10–12	4–8
Alloy steel	8	5–8	12–15	0–4
Brass	0	0	8–12	0–3
Copper	35	35	10–20	13–16
Aluminium	40	40	10–20	15–18

NEGATIVE RAKE CUTTING. Figure 11.1 illustrates a ductile steel being turned, using a carbide lathe tool with a positive rake.

Notice that the cutting force acts on a narrow tool section *AB*, and there is a tendency for the tool to be fractured. This tendency will be even greater if the tool tip has been weakened by cratering.

Figure 11.2 illustrates the same process, using a carbide lathe tool with a negative rake.

Notice that now the cutting force acts on a much wider tool section, with less likelihood of tool breakage. In fact, the tendency now is for the tip to be pressed firmly into its seat. Under these circumstances, higher cutting speeds may be employed, which in effect reduces the cutting force acting on the tool tip. Negative rake cutting increases tool life and gives improved impact resistance. In addition, it is possible to obtain a better surface finish on the workpiece than when positive-rake tools are employed.

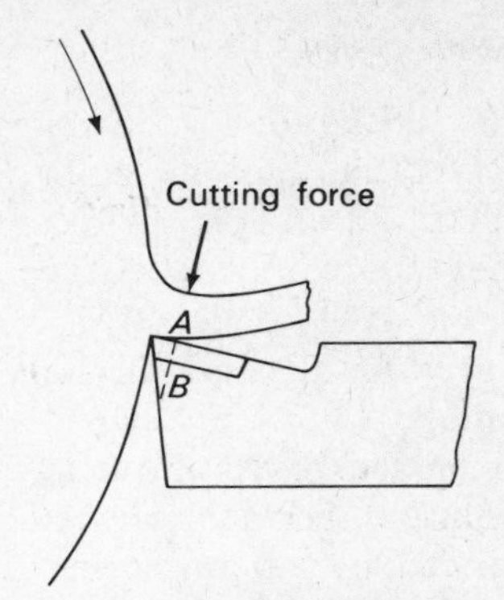

Fig. 11.1 Positive rake cutting

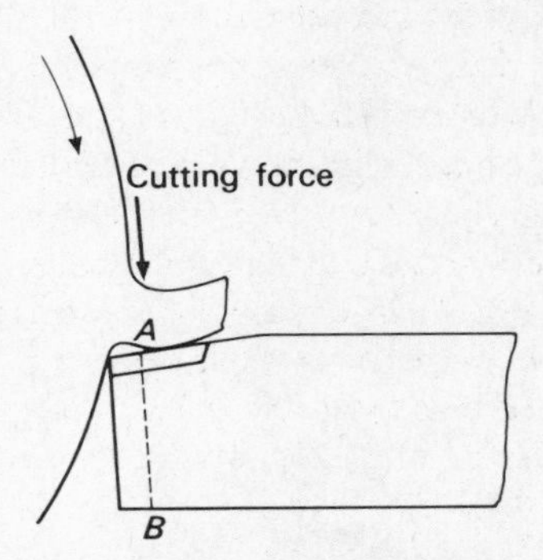

Fig. 11.2 Negative rake cutting

The advantages of negative rate cutting are only obtained when very high cutting speeds are used. This process is, therefore, mainly restricted to machining with carbide and ceramic tools. Also, many workpiece materials, such as cast iron, are not suited to negative rake cutting, and this process is most satisfactorily applied to the machining of ductile steels. For such applications, negative rake angles of between 5° and 10° are commonly employed. Milling cutters, such as inserted-tooth face mills, often operate with negative rake cutting actions and, with these, very high production rates are possible.

11.3 Throw-away tips

The modern cutting-tool tendency is towards using tips of cutting-tool materials, secured in suitable holders. These holders may be designed such that they automatically secure the tips, or bits, at suitable rake angles. This considerably reduces the need for tool grinding and is of particular advantage when very hard cutting tools such as cemented carbides are employed. These materials require super-fine finishes, which can only be obtained by several stages of grinding, followed by burnishing. The burnishing operation is performed using a special plastic wheel impregnated with diamond dust.

(a) HIGH-SPEED STEEL OR STELLITE BIT HOLDER. This type of holder is shown in Fig. 11.3. The slot carrying the bit is machined at an angle to the base, thus automatically providing the required rake and side clearance. The front clearance angle can be repeatedly re-ground until only a very small stump need be thrown away.

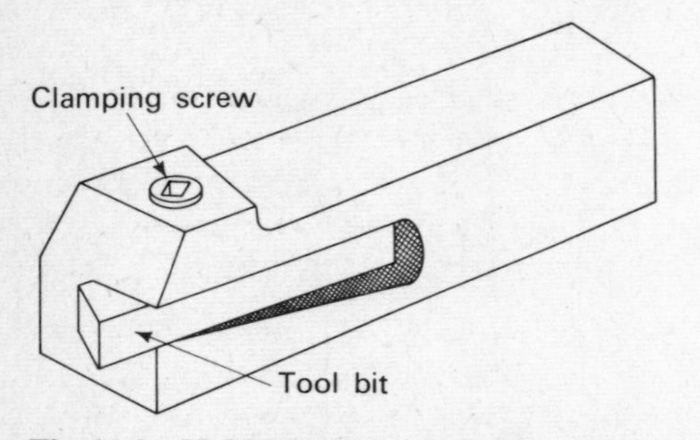

Fig. 11.3 Holder for H.S.S or stellite tool bits

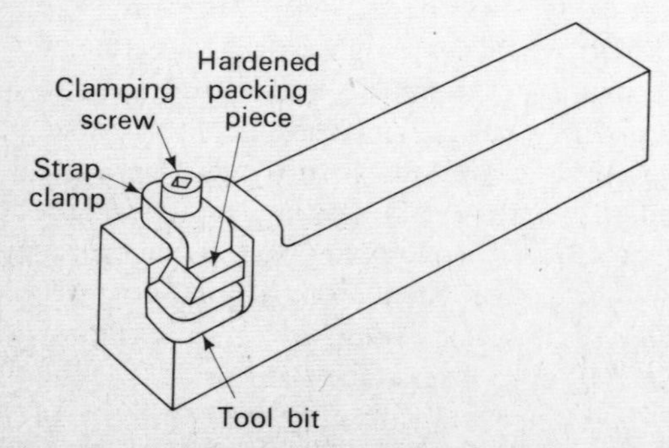

Fig. 11.4 Clamp-held carbide or ceramic tool-bit holder

(b) CARBIDE OR CERAMIC BIT HOLDERS. Short carbide or ceramic bits are produced in various shapes, e.g., square, round, and triangular, and these are secured in holders incorporating clamps or levers, as shown in Figs. 11.4 and 11.5 respectively. With both types, a new cutting edge is obtained by indexing the tip.

The economic advantages of using throw-away tips may be extended, by applying their general principle to a variety of cutting tools. For example, large-diameter face mills used for generating flat surfaces are constructed in the form of a steel body carrying inserted cutting blades. Tapered pins or wedges are employed to secure the blades in slots, which automatically provide the required cutting rake. An inserted-blade face mill is illustrated in Fig. 11.6.

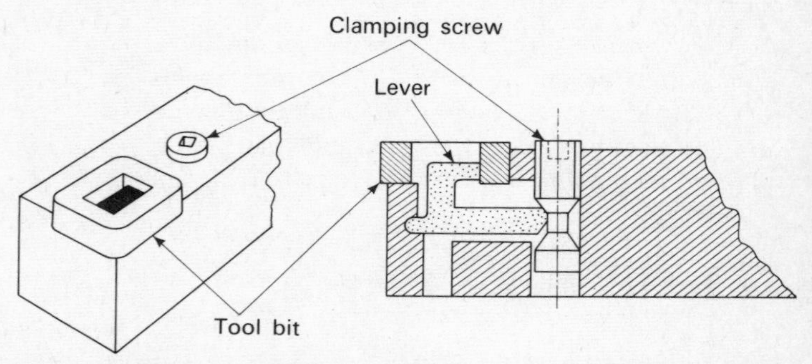

Fig. 11.5 Lever-held carbide or ceramic tool-bit holder

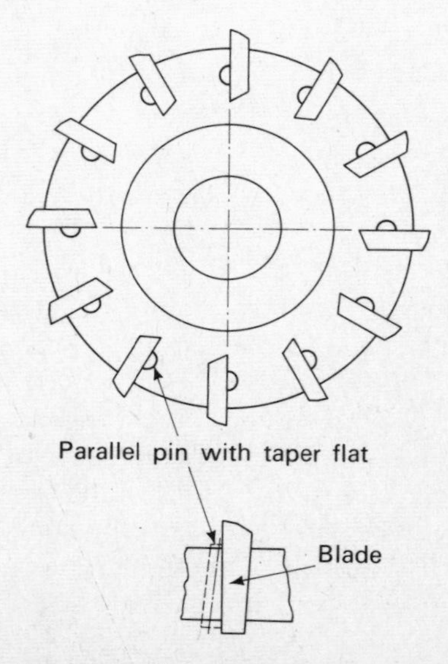

Fig. 11.6 Inserted-blade face mill

11.4 Types of file

Files are manufactured from high-carbon steel and are specified according to three features:

(a) Length
(b) Shape
(c) Cut

The student should already be familiar with common files but brief revision will be given to these before proceeding to special types.

COMMON FILES.

(a) *Length.* The usual sizes manufactured are from approximately 100 mm to 500 mm in steps of 50 mm. Each increase in length is accompanied by an increase in section and spacing between teeth.

(b) *Shape.* Six common shapes are manufactured, these being shown in Fig. 11.7.

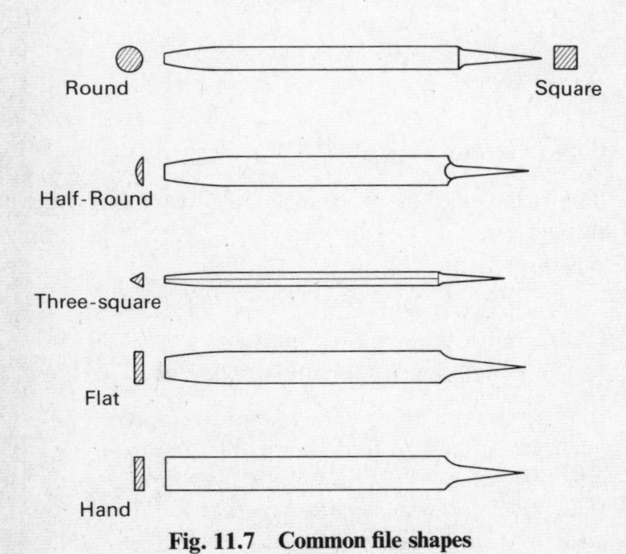

Fig. 11.7 Common file shapes

(c) *Cut.* This indicates the cutting action of a file, and is dependent upon the arrangement of teeth and the spacing between them. The chief grades of cut are shown in Fig. 11.8.

Single-cut files are used to produce a smooth surface finish, whereas for roughing out, double-cut files are employed.

In addition to the grades of cut shown in Fig. 11.8, rasps are also available. These have coarse

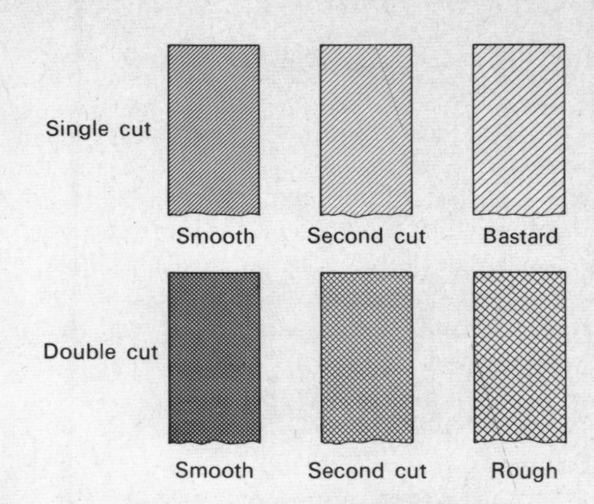

Fig. 11.8 Chief grades of file cuts

teeth, cut with a pointed punch, and are mainly intended for filing soft metals, plastics and wood.

SPECIAL FILES. These are files having less-common shapes than those mentioned previously. They are, however, not rarities, but are intended for more limited ranges of application. Numerous special files are available, the following being typical examples.

(a) *Warding file.* (Fig. 11.9a). This is very similar to the flat file but is only about $1\frac{1}{2}$ mm thick. It is intended to be used only on edge when, for example, filing narrow slots. Warding files are used considerably by machinists, but more especially by locksmiths for filing the ward notches in keys; hence the name.

(b) *Pillar file* (Fig. 11.9b). This is similar to the hand file, but is narrower in proportion to the thickness. Pillar files are double cut, and are useful for filing in narrow apertures.

(c) *Mill files* (Fig. 11.9c). There are several types of mill file, the most usual being similar to the flat file, but parallel on both width and thickness. Mill files may be obtained with either one or both edges rounded, these being called *mill saw* files. These files are frequently used for sharpening metal saws and for radiusing slots.

(d) *Knife file* (Fig. 11.9d). This file is so called because it somewhat resembles the blade of a knife. Knife files are tapered to a thin edge on one side and this may be employed for such purposes as clearing out serrations on shafts.

(e) *Crossing file* (Fig. 11.9e). This file is ideally suited to the production of concave surfaces. Although the half-round file is satisfactory for roughing out, it may not produce a true surface. A crossing file, having a larger radius, will tend to cut only at its outer edge and this will be found to produce a smooth curve without any tendency for the file to 'rock'.

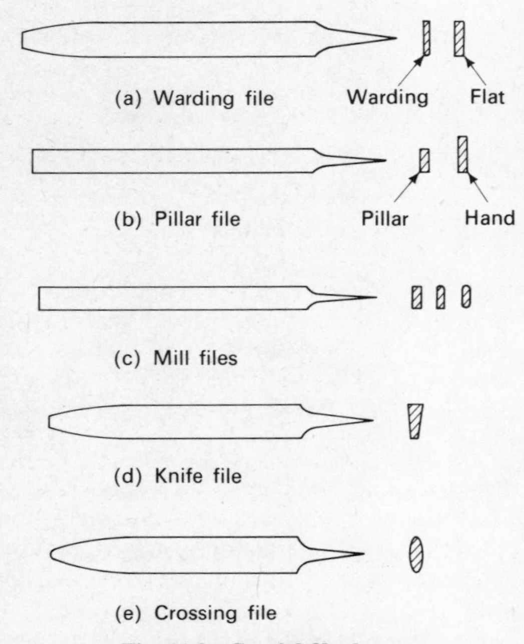

Fig. 11.9 Special file shapes

(f) *Swiss files.* These are used by precision fitters and toolmakers for detail work. Swiss files are manufactured with shapes similar to common files. Special shapes are also available, including those shown in Fig. 11.10.

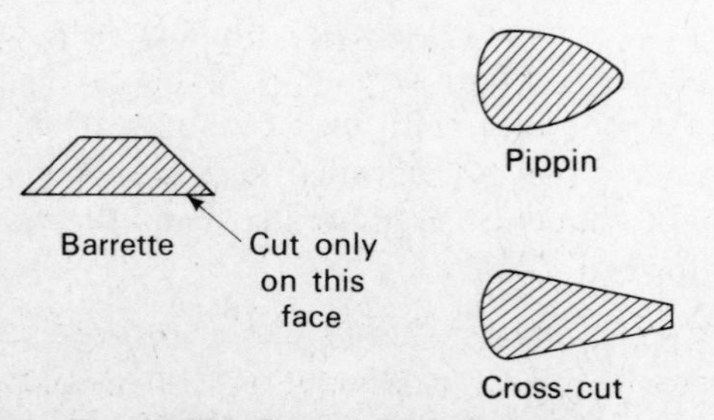

Fig. 11.10 Special shape Swiss files

(g) *Needle files.* These are similar to Swiss files but are formed with handles as shown in Fig. 11.11. Needle files are provided in sets with lengths from approximately 100 mm to 175 mm.

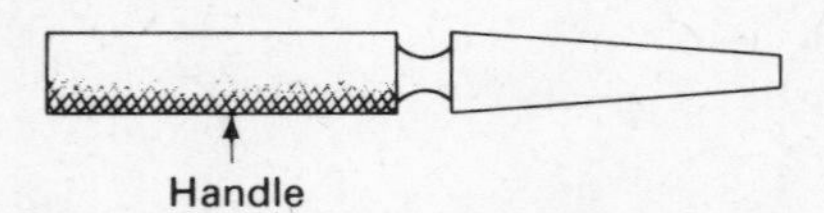

Fig. 11.11 Needle file

Rifflers. These are small double-ended files and are used for 'cleaning up' curved surfaces on jigs, tools and dies. Due to their curved shapes, rifflers are particularly suitable for filing in places not accessible to straight files. A typical riffler is shown in Fig. 11.12.

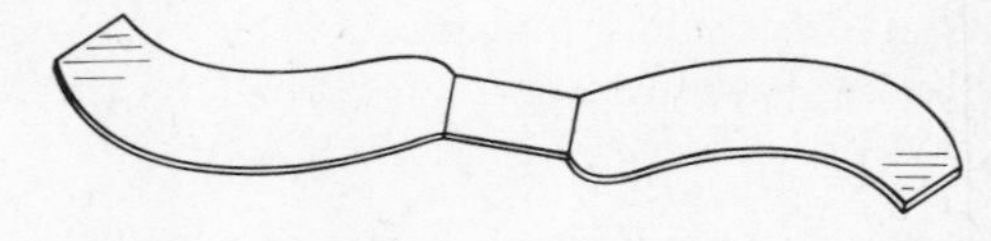

Fig. 11.12 Riffler

11.5 Hand scraping

The surfaces of many components are finished by hand scraping for one or more of the following reasons:

(a) To improve surface accuracy, e.g., flatness
(b) To remove machining marks
(c) To obtain an ornamental appearance.

When flat surfaces are scraped to improve their accuracy, they are first marked to indicate 'high spots' for removal. This entails moving the workpiece over a surface plate which is lightly coated with marking compound such as Prussian blue. Scraping is then repeated until marking reveals two or three high spots on each square centimetre of the entire surface.

Both flat and round bearing surfaces may be scraped to ensure good contact between mating parts. Journal bearings may be marked by lightly coating the shaft with Prussian Blue and rotating this in contact with the bearing surface. Scraped bearings, in addition to providing an accurate fit, also tend to retain surface 'pockets' of oil, this being an advantageous feature.

Flat surfaces may be scraped to provide an ornamental effect, whilst improving accuracy. A common example of this is *snow flaking*, the resulting appearance being similar to a chess board. This is achieved by scraping neighbouring 'squares' at ninety degrees to each other. An alternative appearance consists of rows of crescent-shaped marks. This is called *frosting* and is produced by twisting the wrist during each scraping action.

TYPES OF SCRAPER. Scrapers are manufactured from high-carbon steel, several types being available, including the following:

(a) *Flat scraper* (Fig. 11.13a). This type is used to remove errors in flatness and to produce a smooth surface. It is also suitable for ornamental scraping. The cutting edge is ground slightly convex, to prevent the corners from digging into the work, and this must frequently be honed (sharpened) on an oil stone.

(b) *Hook scraper* (Fig. 11.13b). This type is often preferred for obtaining a smooth, flat surface. It is drawn across the work surface towards the operator and permits very light control.

(c) *Half-round scraper* (Fig. 11.13c). This type is used for finishing internal cylindrical surfaces to a high degree of roundness.

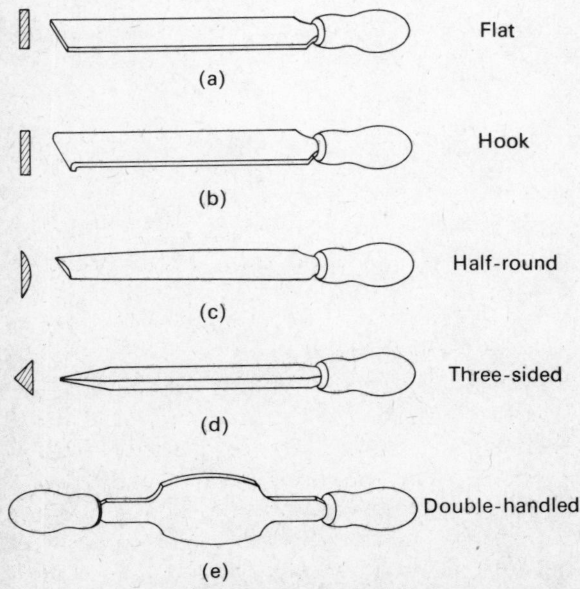

Fig. 11.13 Types of scraper

(d) *Three-sided scraper* (Fig. 11.13d). This type can be used on curved surfaces, but is best suited to producing sharp corners in workpieces.

(e) *Double-handled scraper* (Fig. 11.13e). This type is designed for scraping large journal bearings.

11.6 Lapping

This process consists of 'wearing down' a workpiece surface using an abrasive powder. The resulting surface will have an extremely fine finish and the lapping process can be employed to obtain minute reductions in size. For these reasons, precision tools and gauges, e.g., slip gauges, are finished by lapping.

Various abrasive powders are used, including carborundum and diamond dust, these being mixed with oil or thin grease to form a lapping paste. This is applied to the surface of a lapping plate or *lap*, which is then moved in contact with the workpiece surface. The movement must be random (in no particular direction), to ensure that a flat surface is produced.

The lap is made of a softer material than the workpiece, to ensure that the abrasive will become embedded in the lap rather than the work surface. Grey cast-iron laps are used for lapping hardened steel components, whereas for softer workpieces, copper, brass or lead laps may be employed. The laps must frequently be cleaned during the lapping process, as the paste will become charged with small metal particles from the workpiece.

The lapping process can be applied to both flat and cylindrical surfaces, using suitably shaped laps. Plug and ring gauges may, for example, be lapped to size by the toolmaker. The slowness of this process will be to his advantage when working to very small tolerances. Small circular surface plates, called *toolmaker's flats*, are finished by lapping, their accuracy of flatness being to within 0·0005 mm. Such accuracy may well be much greater than that of the lap employed in their production.

Summary

For the machining of hard materials, and to obtain high production rates, special cutting-tool materials are becoming increasingly used. These include *super high-speed steel*, *stellite*, *cemented carbides*, *ceramics* and *diamonds*. In this order, these materials have increased *hardness*, *red-hardness* and *brittleness*, but decreased *impact resistance*.

Due to the high cost of these materials, tips are often fixed to carbon-steel shanks by such processes as brazing. Alternatively, *throw-away tips* may be secured in suitable holders, which automatically provide appropriate rake and side-clearance angles. These holders reduce tool-grinding costs to a minimum, and ensure that very little cutting-tool material is wasted. Extremely brittle tips, such as cemented carbides and ceramics, are frequently held with negative cutting rakes to provide increased *tool life* and *impact resistance*.

The removal of material by hand processes will often involve the use of *special files*. These are available in many shapes and sizes, typical examples being discussed in section 11.4.

The accuracy and appearance of a workpiece may be improved by *scraping* its surfaces. This process is widely used for finishing both flat and round bearing surfaces, using *scrapers* of the types discussed in section 11.5.

To obtain extremely accurate and smooth surfaces on precision components such as gauges, the *lapping* process may be employed. This involves the use of a *lap* charged with an *abrasive* which 'wears down' the component to its finished form.

Questions

1. List the main properties of the following cutting-tool materials:
   (a) Super high-speed steel
   (b) Stellite
   (c) Cemented carbides
   (d) Ceramics
   (e) Diamonds.
2. Give examples to illustrate the use of each of the cutting materials in question 1.
3. What difficulties are encountered when tungsten carbide tools are used for machining steels?
4. What is meant by *negative-rake cutting*, and what are its advantages?
5. Sketch the side elevation of a carbide-tipped rough-turning tool suitable for cutting mild steel. Indicate the size of the top rake and front clearance angles.
6. Discuss the use of throw-away tips and sketch a suitable holder for securing them.
7. Sketch and name four special files, indicating their uses.
8. Sketch and name three types of scraper and give reasons for their use to finish the surfaces of components.
9. Describe in detail, with the aid of sketches, the process of scraping a 60 mm diameter journal bearing.
10. Describe the lapping process and state examples of its use, indicating the approximate accuracy obtained.

12. Studies associated with material removal

This chapter examines the following topics associated with material removal.

12.1 British Standard tool shapes
12.2 Forces on cutting tools
12.3 Metal-cutting calculations
12.4 Moments on cutting tools
12.5 Surface texture
12.6 Machining symbols

12.1 British Standard tool shapes

Specifications for 'butt-welded, single-point cutting tools' are given in British Standard 1296:1961. The original purpose of this standard was to simplify the variety of sizes and shapes of such tools, and thus enable them to be manufactured more economically.

This British Standard is concerned with lathe, shaper and planer tools, manufactured by butt-welding high-speed steel tips to medium-carbon steel shanks. Suitable shank dimensions are recommended in the Standard, together with tool-point shapes including those shown in Fig. 12.1.

Recommended shapes for *swan-necked* tools are illustrated in Fig. 12.2. These tools are employed for shaping and planing operations, where there is a tendency for the tool to 'spring' during machining. Whereas a straight shank tool is liable to dig into the work surface under the pressure of the cut, a swan-necked tool will 'swing' away from the surface, as shown in Fig. 12.3.

12.2 Forces on cutting tools

In chapter 7 it was noted that during a surfacing operation, a lathe bed must resist two forces, these being shown in Fig. 7.2. The downward pressure of the work acting on the cutting tool is called the *tangential* force. This is so called because its line of

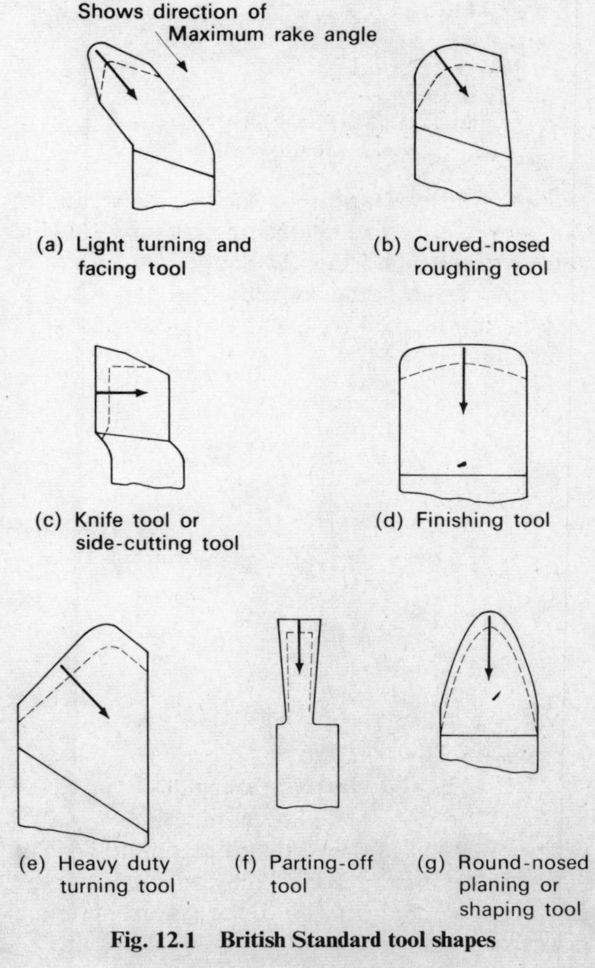

Fig. 12.1 British Standard tool shapes

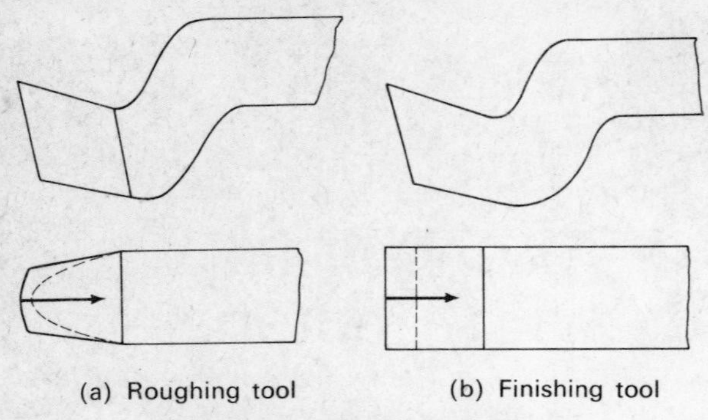

Fig. 12.2 Swan-necked tools

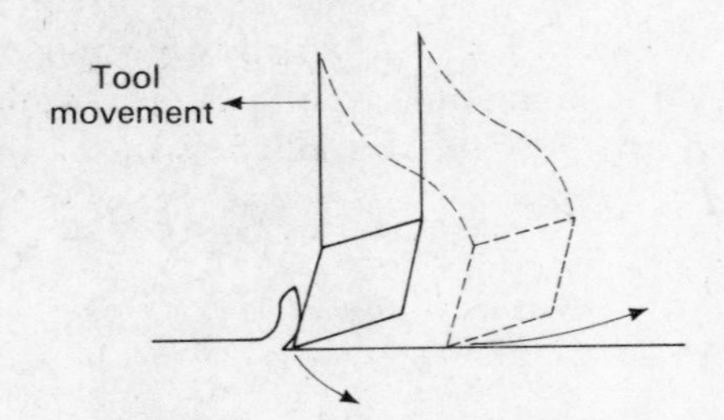

Fig. 12.3 Advantage of swan-necked tools

action forms a tangent to the circumference of the workpiece. The outward pressure on the cutting tool is called the *radial* force.

The net result of the tangential and radial forces may be represented by a single *resultant* force as shown in Fig. 12.4.

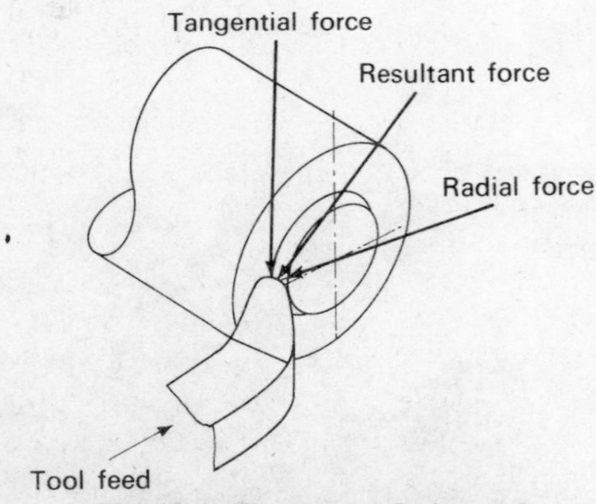

Fig. 12.4 Forces on a facing tool

The direction of the resultant cutting force is of considerable importance, as the design of both the cutting tool and machine tool must provide resistance against its action.

If the values of the tangential and radial forces are known, then the direction and value of the resultant force may be determined.

Example During a surfacing operation the forces on the lathe tool were estimated to be:

Tangential force	1750 N
Radial force	1000 N.

The value and direction of the resultant force may be determined by drawing the given forces to scale and completing a parallelogram, as shown in Fig. 12.5.

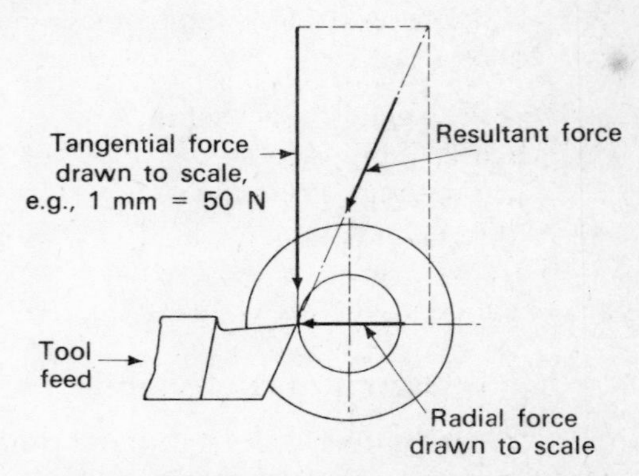

Fig. 12.5 Determination of the resultant force

The resultant cutting force will be found by measurement to have a value of approximately 2000 N at an angle slightly greater than 60° to the horizontal.

During cylindrical turning operations, a third force will act on the cutting tool. This will be a *side* force caused by the feed motion of the tool. The resultant force on a knife tool during a cylindrical turning operation is illustrated in Fig. 12.6.

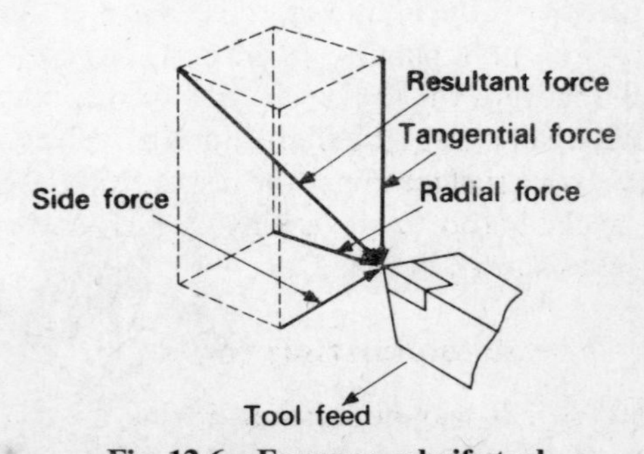

Fig. 12.6 Forces on a knife tool

Similar forces act on cutting tools used in other machining processes. A plain milling cutter is illustrated in Fig. 12.7, the presence of tangential and radial forces being indicated.

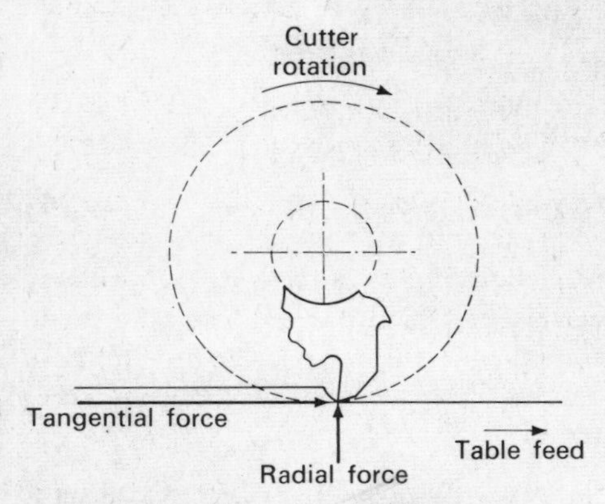

Fig. 12.7 Forces on a milling cutter

Consideration of other milling cutters should enable the student to appreciate that tangential, radial and side forces may all act on cutter teeth.

12.3 Metal-cutting calculations

CUTTING SPEED. This is the speed at which material is removed from the surface of a workpiece during a cutting operation. All materials have ideal cutting speeds, these being quoted in either m/min or m/s. When, for example, mild steel is machined, using a high-speed steel tool or cutter, the ideal cutting speed is 30 m/min (0·5 m/s). This means that if the material was removed in the form of a continuous chip, this would be thirty metres in length after cutting for one minute.

Table 12.1 Cutting speeds

Material to be cut	Cutting speed (m/min), using			
	High-speed steel	Stellite	Cemented carbides: Roughing	Cemented carbides: Finishing
Mild steel	30	50	90	180
Grey cast iron	25	30	50	120
Alloy steels	15	25	40	100
Brass	50	80	180	350
Copper	100	100	180	350
Aluminium	150	200	150	250
Plastics	180	180	180	300

Typical cutting speeds for various workpieces and cutting-tool materials are quoted in Table 12.1.

The figures quoted in Table 12.1 are only general guides, and must be modified to suit cutting conditions. When, for example, fragile milling cutters are used, these values must be lowered. Similarly, when negative-rake cemented-carbide tools are used, the cutting speeds employed may be two or three times greater than those shown:

SPINDLE SPEED. This is the speed at which a machine spindle rotates during a cutting operation, and is quoted in either rev/min or rev/s. Given the cutting speed for a certain machining operation, it is possible to calculate the required spindle speed from the formula:

$$N = \frac{1000 \times S}{\pi \times D},$$

where

N = spindle speed rev/min or rev/s

S = cutting speed m/min or m/s

π = 22/7 or 3·142

D = diameter rotating mm.

Note that, for lathework, the diameter rotating will be that of the work, whereas for milling and drilling it will be that of the cutter.

Example Calculate a suitable spindle speed for finish turning a 35 mm diameter, mild-steel shaft, using a cemented-carbide cutting tool.

$$N = \frac{1000 \times S}{\pi \times D}.$$

From Table 12.1, S = 180 m/min.

$$\therefore \quad N = \frac{1000 \times 180}{(22/7) \times 35} = \frac{1000 \times 180 \times 7}{22 \times 35}$$

$$= \frac{18\,000}{11} = N = 1636 \text{ rev/min.}$$

(to nearest whole number)

$\therefore$ suitable spindle speed = closest available to 1636 rev/min.

STROKE SPEED. This is the speed at which the ram moves during shaping operations, and is quoted in strokes per minute. Given the cutting speed for a shaping operation, it is possible to calculate the stroke speed from the formula:

$$N = \frac{1000 \times S}{L} \times T_f$$

where

N = stroke speed strokes/min
S = cutting speed m/min
L = stroke length mm
T_f = fractional time spent in cutting.

Example Calculate a suitable stroke speed for a shaping operation, given that:

(a) the cutting speed used is 24 m/min
(b) the stroke length is 300 mm
(c) the cutting–return time ratio is 5:3.

$$N = \frac{1000 \times S}{L} \times T_f$$

$$= \frac{1000 \times 24}{300} \times T_f$$

Note that the cutting–return time ratio is 5:3. This means that if the forward or cutting stroke is imagined to take five seconds, then three seconds will be spent on the return stroke. It follows, therefore, that the cutting stroke occupies five-eighths of the total time and it is this period only that the cutting-speed information applies to.

$$\therefore \quad N = \frac{1000 \times 24}{300} \times \frac{5}{8}$$

$$80 \times 5/8 = N = 50$$

$\therefore$ suitable stroke speed = 50 strokes/min.

FEED. This is the linear traverse motion employed during a cutting operation. For lathework, the feed is applied to the tool, and is quoted in mm per revolution of the spindle.

During milling operations, the feed is applied to the table and workpiece. In the case of milling cutters, the feed per revolution would not give a clear indication of the work done by each cutter tooth. For this reason milling feeds are generally expressed in mm per tooth, typical values being given in Table 12.2 for milling steel.

Table 12.2 Milling feeds

Type of cutter	Feed/tooth (mm), using	
	High-speed steel	Cemented carbides
Slab mill	0·1–0·25	0·2–1·3
Face mill	0·1–0·4	0·2–0·5
Slotting cutter	0·1	0·1–0·25
Saw	0·05	0·05–0·15
End mill	0·05–0·25	0·05–0·25

Given the number of teeth in the milling cutter, the feed per tooth, and spindle speed, it is possible to calculate the table feed rate in mm/min. This gives a clear indication of production rates and is calculated from the formula:

Table feed rate (mm/min) = number of teeth in cutter × feed per tooth (mm) × spindle speed (rev/min)

Example A certain high-speed steel slotting cutter is 120 mm diameter and has 16 teeth. If it is used for cutting mild steel, calculate a suitable table feed rate in mm/min. Now from Table 12.1,

suitable cutting speed = 30 m/min

and from Table 12.2,

suitable feed per tooth = 0·1 mm.

Therefore, from the formula:

table feed rate = 16 × 0·1 × spindle speed,

the spindle speed must now be calculated:

$$N = \frac{1000 \times S}{\pi \times D} = \frac{1000 \times 30}{(22/7) \times 120}$$

$$= \frac{1000 \times 30 \times 7}{22 \times 120} = \frac{250 \times 7}{22}$$

$\therefore$ spindle speed = 80 rev/min (to nearest whole number).

Finally,

table feed rate = 16 × 0·1 × 80

$\therefore$ suitable table feed rate = 128 mm/min.

CUTTING TIMES. The cutting time required for machining operations may be calculated as follows:

(a) *Turning and drilling*

$$\underset{\text{(min or s)}}{\text{Cutting time}} = \frac{L}{N \times F}$$

where

L = length of cut	mm
N = spindle speed	rev/min or rev/s
F = feed rate	mm/rev.

Example A bar is turned 20 mm diameter, for a length of 150 mm, using a spindle speed of 480 rev/min and a feed of 0·25 mm/rev. Calculate the cutting time of the operation.

$$\text{Cutting time} = \frac{L}{N \times F} = \frac{150}{480 \times 0{\cdot}25}$$

$$= \frac{150}{480 \times 1/4} = \frac{150 \times 4}{480}$$

$$= \frac{600}{480} = \frac{5}{4}$$

$\therefore$ $\underline{\text{cutting time} = 1\tfrac{1}{4}\text{ min.}}$

(b) *Milling.*

$$\underset{\text{(min or s)}}{\text{Cutting time}} = \frac{L}{F}$$

where

L = length of cut	mm
F = table feed rate	mm/min or mm/s.

Example A slab mill is used to mill a plane surface 300 mm in length, at a table feed rate of 125 mm/min. Calculate the cutting time of the operation.

$$\text{Cutting time} = \frac{L}{F} = \frac{300}{125} = \frac{12}{5}$$

$\therefore$ $\underline{\text{cutting time} = 2\tfrac{2}{5}\text{ min.}}$

(c) *Shaping.*

$$\underset{\text{(min)}}{\text{Cutting time}} = \frac{W}{N \times F},$$

where

W = width of cut	mm
N = stroke speed	stroke/min
F = feed rate	mm/stroke.

Example A plane surface 160 mm wide is shaped at 40 strokes/min, using a feed rate of 2 mm/stroke. Calculate the cutting time of the operation.

$$\text{Cutting time} = \frac{W}{N \times F} = \frac{160}{40 \times 2}$$

$\therefore$ $\underline{\text{cutting time} = 2\text{ min.}}$

METAL REMOVAL RATE. This indicates the volume of metal removed in a given time, and may be calculated from the formula:

$R = V/T$

where

R = metal removal rate	mm^3/min or cm^3/h
V = volume of metal removed	mm^3 or cm^3
T = cutting time	min or h.

Example The top surface of a block of metal, measuring 125 mm × 50 mm × 60 mm deep, is shaped using a depth of cut of 4 mm. If the cutting time of the operation is $2\tfrac{1}{2}$ min, calculate the metal removal rate in mm^3/min.

$$R = V/T$$

and

$$V = \text{length} \times \text{width} \times \text{depth of cut}$$

$$= 125 \times 50 \times 4$$

$$\therefore \quad R = \frac{125 \times 50 \times 4}{2\tfrac{1}{2}} = \frac{125 \times 50 \times 4 \times 2}{5}$$

$\therefore$ $\underline{\text{metal removal rate} = 10\,000\text{ mm}^3\text{/min.}}$

Example Using information given in the previous example, calculate the metal removal rate in cm^3/h.

$$R = V/T,$$

where R is in cm^3/h if V is in cm^3 and T is in h.

Now since 10 mm = 1 cm,

depth of cut = 4 mm = 4/10 cm

length = 125 mm = 125/10 cm

width = 50 mm = 50/10 cm.

Also,

$$2\tfrac{1}{2}\text{ min} = 2\tfrac{1}{2}/60\text{ h} = 5/120\text{ h}$$

$$\therefore \quad R = \frac{(125/10) \times (50/10) \times (4/10)}{(5/120)}$$

$$= \frac{125 \times 50 \times 4 \times 120}{10 \times 10 \times 10 \times 5}$$

$$\therefore \quad \underline{\text{metal removal rate} = 600\text{ cm}^3/\text{h}.}$$

12.4 Moments on cutting tools

In section 6.1, it was seen that the effect of a force could be magnified by leverage. The application of a force at some distance from a fulcrum or axis of rotation, was found to result in a *turning moment* or *moment of force*, its value being calculated from the formula

$$\text{moment of force} = \text{force} \times \text{perpendicular distance from fulcrum.}$$

In section 12.2, it was seen that three forces may act on a cutting tool during a machining process. During a turning operation, the tangential force will tend to tilt the tool point downwards. This effect will be magnified if the tool protrudes an excessive distance from the toolpost, as shown in Fig. 12.8.

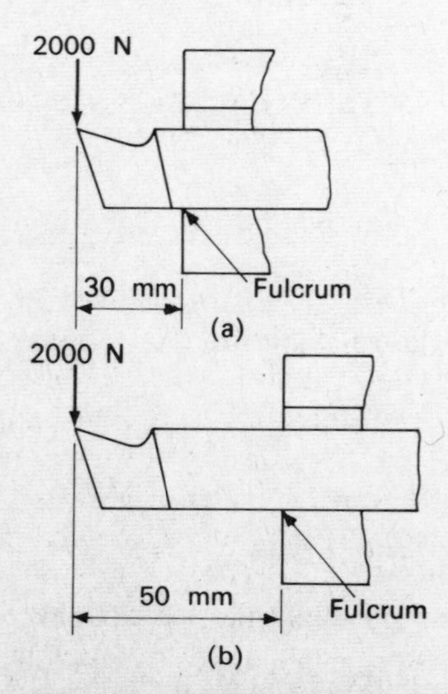

Fig. 12.8 Moment of force on lathe tools

Referring to Fig. 12.8(a),

$$\text{moment of force} = 2000 \times 30 = 60\,000\text{ Nmm} = \underline{60\text{ Nm.}}$$

Referring to Fig. 12.8(b),

$$\text{moment of force} = 2000 \times 50 = 100\,000\text{ Nmm} = \underline{100\text{ Nm.}}$$

Note that, in each case, the tool point tends to turn in an anticlockwise direction. If tilting is to be prevented, then a force must be applied which tends to turn the tool in a clockwise direction. This force is provided by the securing screws in the tool post.

12.5 Surface texture

A machined surface may appear smooth to the naked eye and be said to have a good finish. If, however, it is viewed under a microscope, surface roughness will be visible. The degree of roughness will vary for different machining processes, and this may be measured with the aid of special instruments.

These instruments incorporate a highly sensitive stylus, which rides up and down in the roughness grooves as it crosses a machined surface. The movement of the stylus is recorded electrically within the instrument, and a graph is automatically produced which illustrates the surface texture greatly magnified. A centre-line is then drawn on the graph, as shown in Fig. 12.9, such

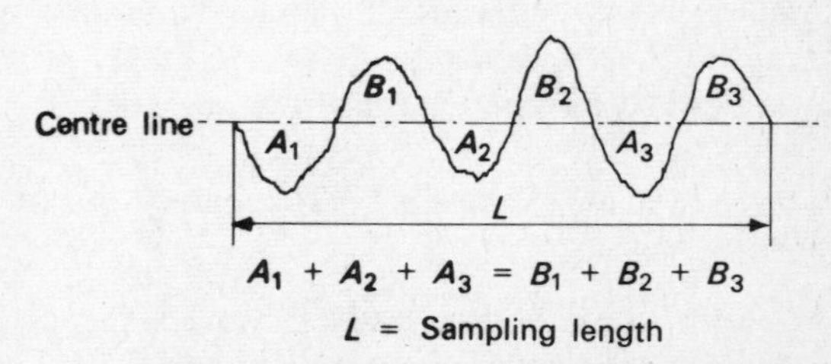

Fig. 12.9 Surface texture graph

that the sum of the areas above the line is equal to the sum of the areas below it. An index, called the centre-line average (C.L.A.), is then calculated from the formula

$$\text{C.L.A. index number} = \frac{\text{sum of all areas} \times K}{L},$$

where K is a constant depending upon the magnification of the graph and the units of measurement employed.

Many instruments used for measuring surface textures give a direct reading of the C.L.A. index number, by means of a pointer on a scale. This C.L.A. index is expressed in units of one-millionth of a metre, or micrometre (μ). Typical values of surface finish are given in Table 12.3.

Table 12.3 Surface finishes

Machining process	Surface texture obtained C.L.A. μ
Rough turning Rough boring Rough milling Rough planing Rough slotting	3·2–12·5
Finish turning Finish boring Finish milling Finish planing Finish slotting	0·8–3·2
Drilling	1·6–12·5
Reaming	0·4–3·2
Grinding	0·1–0·4
Honing	0·1–0·8
Lapping	0·025–0·4

12.6 Machining symbols

An engineering drawing is used to convey to the craftsman such information as:

(a) The finished shape of the workpiece
(b) The finished dimensions of the workpiece
(c) The degree of accuracy required on finished dimensions.

The form of material supplied, however, may be such that not all surfaces require machining. It may only be necessary, for example, to machine location faces on a casting. For this reason, surfaces requiring machining should be indicated on the drawing by a machining symbol, as shown in Fig. 12.10.

Where all the surfaces of a component are to be machined, a general note should be included on the drawing, such as that shown in Fig. 12.11(a).

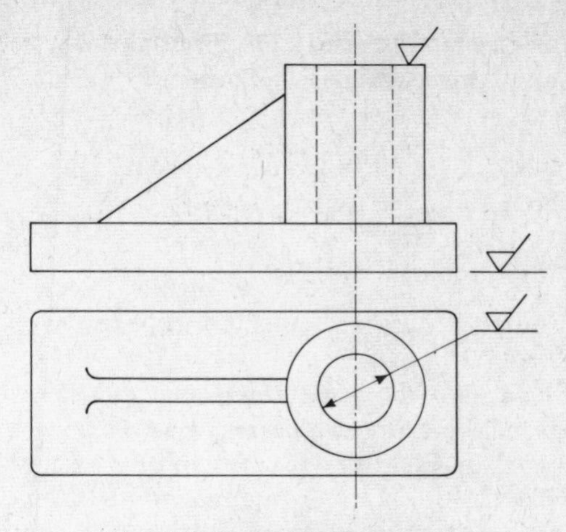

Fig. 12.10 Machining symbols

Where the permissible roughness of the finished surface is of considerable importance, this may be indicated by including the C.L.A. index number with the machining symbol.

When this quality of surface finish can be obtained only by one particular process, e.g., lapping, then this process should also be indicated on the machining symbol as shown in Fig. 12.11(b).

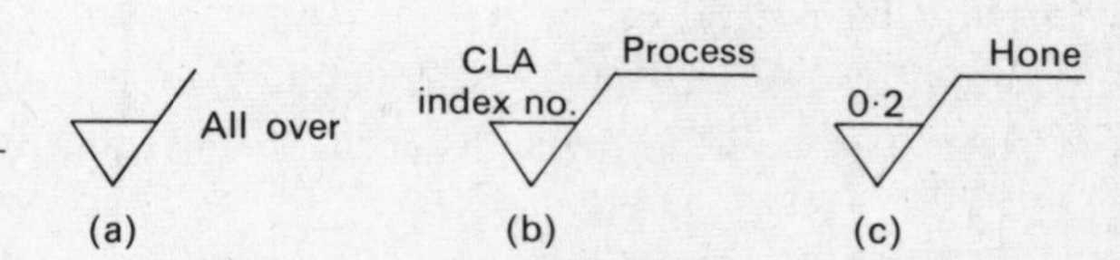

Fig. 12.11 Machining symbols: (a) general note for machining; (b) detail machining symbol; (c) example illustrating (b)

Summary

Standardization has enabled more economic manufacture of high-speed steel tips and suitable medium-carbon steel shanks, typical lathe, shaper and planer tool shapes being illustrated in section 12.1.

Three forces may act on a cutting tool point during a machining process, these being called *tangential*, *radial*, and *side* forces. The net result of these forces is called the *resultant* force, and cutting tools and machine tools must be designed to withstand its action.

Various metal-cutting calculations may be required in the machine shop, these being concerned with:

(a) Cutting, spindle and stroke speeds
(b) Tool and table feeds
(c) Cutting times
(d) Metal removal rates.

Examples of such calculations are given in section 12.3.

A *moment of force* or *turning moment* is the effect produced by a force acting at some distance from the pivot, or *fulcrum*. The moment of force tending to deflect a tool point during machining should be kept to a minimum by satisfactory tool mounting, as indicated in section 12.4.

A degree of roughness is present in all components and workpiece surfaces. This may be measured, using special instruments which record a roughness index, in micrometres, called the *centre-line average* (C.L.A.).

Surfaces requiring machining should be indicated on an engineering drawing by means of symbols. *Machining symbols* may include information indicating the *surface texture* required and a suitable machining process.

Questions

1. Sketch and name four British Standard tool-point shapes.
2. Sketch and name the three forces acting on the tool point during a cylindrical turning operation. Also, indicate the resultant force and explain its importance.
3. Calculate a suitable spindle speed in rev/min for drilling a 14 mm diameter hole in mild steel, using a carbide-tipped drill. Assume a suitable cutting speed of 110m/min.
4. Calculate a suitable stroke speed for a shaping operation, given that:
   (a) The cutting speed used is 24 m/min
   (b) The stroke length is 200 mm
   (c) The cutting–return time ratio is 2:1.
5. A milling cutter of 100 mm diameter has 12 teeth, and is used at a cutting speed of 22 m/min. Calculate the feed per tooth, if it takes 1 min to cut a length of 126 mm.
6. Calculate the time required to finish turn a 35 mm diameter bar for a length of 170 mm using a feed rate of 0·2 mm/rev and a cutting speed of 33 m/min.
7. A bar is turned down from 50 mm dia to 40 mm dia for a length of 140 mm in 4 minutes. Calculate the metal removal rate in cm^3/h.
8. What does a C.L.A. index number indicate, and how is it determined?
9. Sketch two machining symbols, and explain when they would be used.

Answers

3. 2500 rev/min
4. 80 strokes/min
5. 0·15 mm/tooth
6. $2\frac{5}{6}$ min
7. 1485 cm^3/h.

13. Planning

Before the manufacture of any workpiece is started, a suitable method of production should be completely planned. Various aspects of planning have been mentioned during the course of this volume, and these may now be combined for further consideration.

This chapter examines the following topics:

13.1 Planning considerations

A suitable method of manufacturing any workpiece may be decided by studying the information provided on an engineering drawing. At the same time, however, it will be necessary to bear in mind the production equipment and machinery available.

The following factors will strongly influence the planning of production methods:

(a) Production quantity
(b) Material and form of supply
(c) Workpiece size and shape
(d) Accuracy and surface finish
(e) Cost
(f) Safety.

(a) PRODUCTION QUANTITY. Workpieces may be manufactured in mass quantities, batches or as single pieces. The production of mass quantities or large batches of components will warrant the attention of a planning department. Planners in such a department will be able to decide the necessity for obtaining special production tools, equipment and machinery. The craftsman, however, may be left entirely responsible for deciding methods of producing small batches of components or single pieces. It is, therefore, this scale of production that will be considered at present.

(b) MATERIAL AND FORM OF SUPPLY. Different workpiece materials will necessitate different approaches to the planning of manufacture. For example, soft materials such as aluminium will permit much higher metal removal rates to be employed than when machining hard steels. The form of material supply may also influence planning. Castings, for example, will often require complete marking out and special work-holding set-ups.

(c) WORKPIECE SIZE AND SHAPE. The method of holding a workpiece during manufacture will largely depend on its size and shape. Very large components present obvious work-holding difficulties, which may well dictate the most suitable method of manufacture.

(d) ACCURACY AND SURFACE FINISH. The degree of accuracy and surface finish required will often indicate the need for particular processes during production. Grinding, for example, may be necessary as a finishing operation, for producing high-quality work.

(e) COST. The cost of any manufactured article must be kept to a reasonable level, to ensure employment for all those concerned with its production. It is the responsibility of everyone, from the production manager to the craftsman himself, to ensure that work is produced by the most economical methods available. The craft student should therefore be encouraged to consider all alternative methods of producing an article, to decide upon the simplest and quickest satisfactory method of production. The method chosen should not offer greater accuracy than that required for the component, in order to prevent unnecessarily high production costs.

(f) SAFETY. Although mentioned last, safety is, and must always be, the prime factor governing all methods of production. Each of the previously mentioned planning factors must always be viewed with safety in mind and this must be given priority over all other considerations.

13.2 Operation sequences

The planning of a suitable method of production may be no more than a quick mental process, when manufacturing articles requiring few operations. For more complex work, however, an *operation sheet* may well be desirable. This sheet should list a sequence of operations, together with all tools and equipment required. In addition, it should indicate datums from which settings and operations should be carried out.

Thorough consideration is given to preparation of operation sheets in Volume 2. Therefore, at this stage, only one simple operation sequence will be planned, this being for the manufacture of the single component shown in Fig. 13.1.

It should be apparent to the student that the manufacture of this component will require turning operations at a minimum of two settings, followed by milling. Work holding, during the milling process, could satisfactorily be achieved by securing the component on a plain mandrel of the type shown in Fig. 5.11. This could then be mounted between the centres of a dividing head and tailstock. It would, therefore, be convenient to complete the turning operations using this mandrel, thus ensuring concentricity between the component's features. A suitable sequence of operations would therefore be as in Table 13.1.

Table 13.1 Sequence of operations

Operation number	Description of operation	Equipment and tools required
1st setting		Centre lathe
1	Grip bar in three-jaw chuck and face one end	Facing tool
2	Reverse bar and face to length	Facing tool, vernier caliper
3	Centre drill, drill and bore 25·04/25·00 mm dia	Centre drill, 8 mm dia drill, 22 mm dia drill, boring tool, plug gauge
4	Turn 50 mm dia, to produce a shoulder width of 25 mm	Roughing tool, knife tool, 25–50 mm dia outside micrometer, vernier caliper
5	Form undercut	Undercutting tool

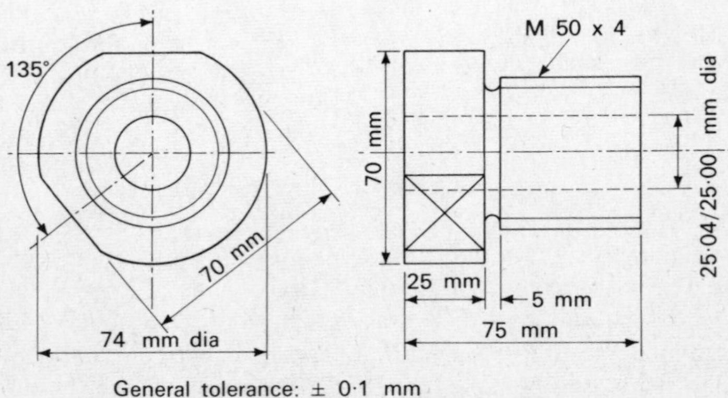

Fig. 13.1 Component

Table 13.1 (contd.)

Operation number	Description of operation	Equipment and tools required
6	Screwcut	Screwcutting tool, screw thread gauges
2nd setting		Centre lathe
7	Mount component on plain mandrel between centres, and turn 74 mm dia	Plain mandrel finishing tool, 50–75 mm dia outside micrometer
3rd setting		Vertical milling machine
8	With component mounted on plain mandrel between centres, mill one flat	Plain mandrel, dividing head and tailstock, 50 mm dia shell end mill
9	Index dividing head and mill second flat *N.B.*, index $= \text{angle} \times 1/9$ $= 135° \times 1/9$ = 15 turns of handle	Plain mandrel, Dividing head and tailstock, 50 mm dia shell end mill
10	Remove all burrs and sharp edges	

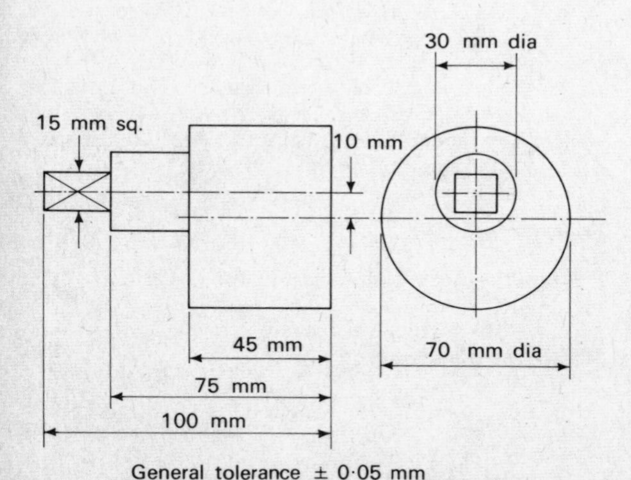

Fig. 13.2 Component

Summary

A suitable method for manufacturing a workpiece may be planned by studying an engineering drawing, whilst bearing in mind the production equipment and machinery available. Important factors to be considered during planning include: *quantity* of workpieces required, workpiece *material*, *size* and *shape*, *accuracy* and *surface finish* required, *economics*, and *safety*.

When planning the production of complex workpieces, the preparation of an *operation sheet* may well prove helpful. This should indicate a suitable *sequence of operations* together with all relevant information, such as tools, equipment and settings required.

Questions

1. Discuss, in your own words, the main factors to be considered when planning the production of workpieces.
2. Prepare operation sheets for the manufacture of the components shown in Figs. 13.2 and 13.3. In each case list a suitable sequence of operations, together with tools and equipment required.

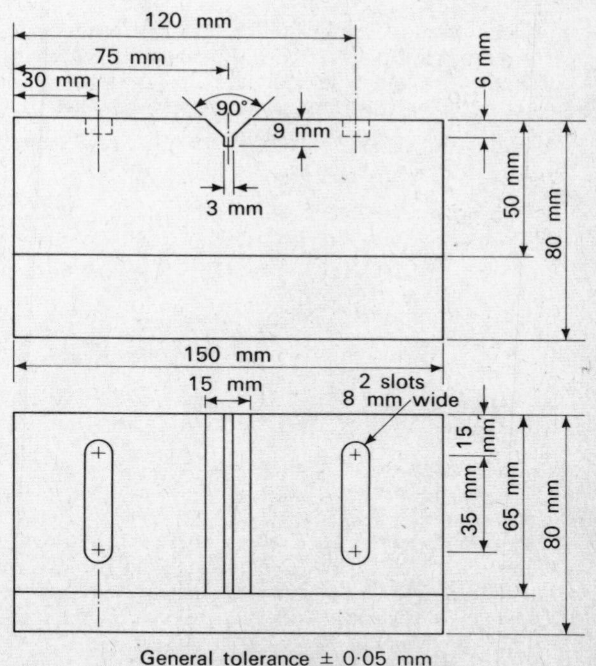

Fig. 13.3 Component

Revision

Test Paper A

This paper is divided into two sections.
Answer all 10 questions in section A } Time 2 hours.
Answer any 2 questions from section B }

Section A

1. Explain the meaning of the terms *forming* and *generating*, when applied to machining operations.
2. Calculate a suitable spindle speed in rev/s for finish turning a mild-steel shaft of 53 mm diameter, using a cutting speed of 0·5 m/s.
3. Describe a turning operation requiring the use of a fixed steady.
4. A force of 120 N applied to a vice handle produces a turning moment of 15 Nm. Calculate the effective length of the handle in mm.
5. State the metric micrometer reading shown in Fig. A.1.

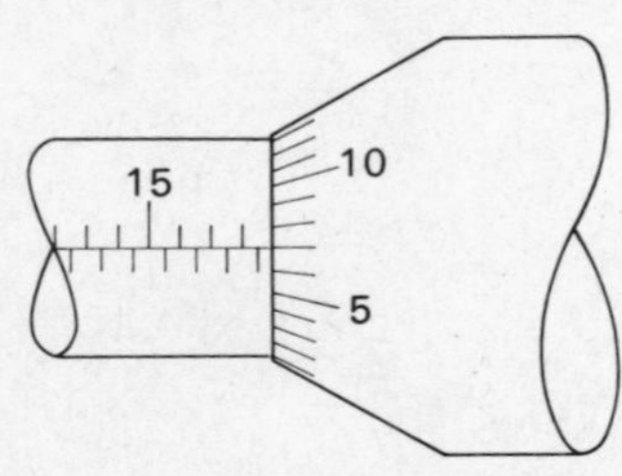

Fig. A.1

6. Describe one workshop application of each of the following location methods:
   (a) Tenons (b) Dowels.
7. An aluminium bar has a length of 600 mm at a temperature of 20°C. If the temperature of the bar is raised to 170°C, calculate its increase in length. (Given: Coefficient of linear expansion of aluminium $= 2{\cdot}4 \times 10^{-5}/°C$.)
8. Name three heat-treatment processes, and state reasons for their being carried out.
9. The net pull in a lathe-spindle driving belt is 400 N and the cutting force is 6 kN. Calculate the force ratio of the driving mechanism.
10. List four reasons why cast iron is widely used for the manufacture of machine-tool frames.

Section B

11. A 30 mm diameter mild steel spindle, 200 mm in length, requires case hardening. Describe in detail:
    (a) How the spindle could be 'pack carburized'
    (b) The subsequent heat treatment required.
12. An 85 mm dia milling cutter has 14 teeth and is used for machining mild steel at a cutting speed of 0·4 m/s with a feed per tooth of 0·1 mm. Calculate:
    (a) suitable spindle speed in rev/s
    (b) suitable table feed rate in mm/s.
13. (a) State two advantages of *limit gauging* over direct measurement.
    (b) Compare the merits of *progressive* and *double-ended* limit gauges.
    (c) Sketch suitable gauges for checking 20·00/20·04 mm diameter shafts, giving reasons for your choice.
14. A shaft is machined with three stepped diameters:
    (a) Explain how concentricity between the stepped diameters could be checked
    (b) What means would be suitable for checking each diameter if the tolerance given was ±0·01 mm?
    (c) Why should the step lengths be dimensioned and measured from one end of the shaft?

Test paper B

This paper is divided into two sections.

Answer all 10 questions in Section A } Time 2 hours.
Answer any 2 questions from Section B }

Section A

1. Briefly describe the process of *lapping* and state a typical application.
2. The gear wheels of a compound gear train are:
   1st driver 30 teeth 1st driven 40 teeth.
   2nd driver 50 teeth 2nd driven 90 teeth.
   If the input shaft rotates at 6 rev/s, calculate the speed of the output shaft.

3. Give three examples where friction is of advantage to the engineer, and three where it is a disadvantage.
4. The limits of size on a shaft and hole are as follows:

	Hole	Shaft
Maximum diameter	25·01 mm	24·97 mm
Minimum diameter	25·00 mm	24·96 mm.

   Calculate the extremes of fit possible, and state the type of fit in each case.
5. Compare the merits of high-speed steel and cemented carbides, as cutting tool materials.
6. The cross-slide of a centre lathe is controlled by means of a lead screw having a single-start thread of 4 mm pitch. Calculate the number of divisions on the indexing dial needed to provide 0·02 mm increments of movement.
7. Make outline drawings of two types of mandrel and describe the type of work for which each would be ideally suited.
8. Outline three possible causes of chatter in one of the following machining operations:
   (a) milling a flat surface.
   (b) reaming a bore.
9. The flange shown in Fig. B.1 is to be faced in a centre lathe. Calculate the time required for this operation, if a spindle speed of 2 rev/s and feed rate of 0·1 mm/rev are used.
10. Explain how alignment between lathe centres may be checked and adjusted.

Section B

11. (a) Describe in detail three methods of turning a tapered component on a centre lathe.
    (b) State which method you would use for turning the taper on the component shown in Fig. B.2.
12. (a) Six equally spaced holes are drilled around a pitch circle of 80 mm diameter. Calculate the centre distance between adjacent holes.
    (b) Calculate the included angle of the taper on the component shown in Fig. B.2.
13. A small surface plate is to be manufactured from a rough casting measuring approximately 300 mm × 200 mm × 50 mm deep. Describe in detail how you would machine and scrape the casting.
14. List a sequence of operations for manufacturing the component shown in Fig. B.3.

Fig. B.1

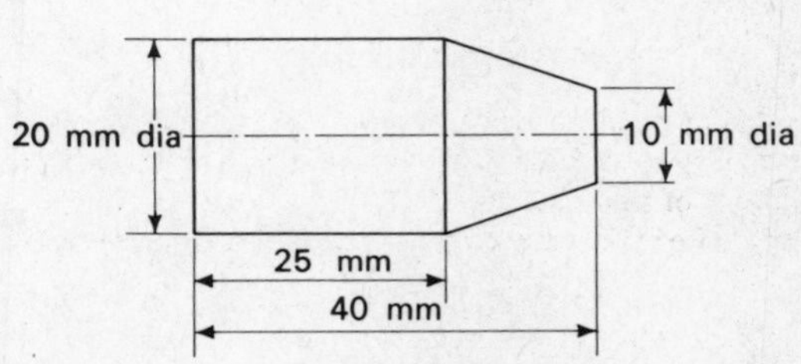

Fig. B.2

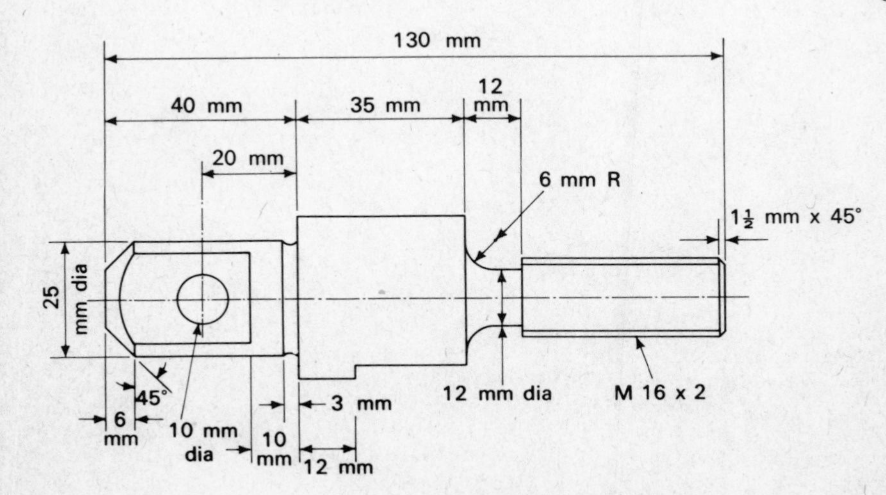

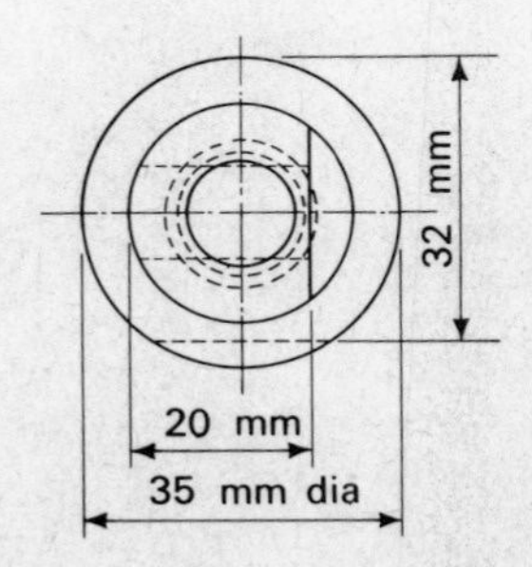

Material: 40 mm dia M.S.
Gen Tol: ± 0·04 mm

Fig. B.3

Answers

Test Paper A

Section A

2. 3 rev/s
4. 125 mm
5. 18·57 mm
7. 2·16 mm
9. 15:1

Section B

12. (a) 1·5 rev/s (b) 2·1 mm/s.

Test Paper B

Section A

2. 2·5 rev/s
4. 0·05 mm clearance
0·03 mm clearance
6. 200
9. 350 s

Section B

12. (a) 40 mm (b) 36° 50′.

Index